United States
Environmental
Protection Agency

Office of Water

NPDES Best Management Practices Manual

Government Institutes, Inc.

Government Institutes, Inc.,
4 Research Place, Suite 200, Rockville, Maryland 20850.

Published April 1995.

99 98 97 96 95 5 4 3 2 1

This guidance manual was prepared by the Office of Water at the U.S. Environmental Protection Agency in October 1993. Government Institutes determined that it contained information of interest to those outside EPA, so we are reproducing this public domain material in order to serve those interested.

The purpose of this manual is to provide guidance to NPDES permittees in the development of best management practices (BMP) for their facilities. It is also useful to permit writers and inspectors charged with evaluating the adequacy of BMP plans. Additionally, it promotes the integration of pollution prevention concepts and practices in BMP plans.

ISBN: 0-86587-466-2

Printed in the United States of America

TABLE OF CONTENTS

1. INTRODUCTION TO BEST MANAGEMENT PRACTICES 1-1
1.1 PURPOSE OF THIS MANUAL 1-1
1.2 BACKGROUND OF NPDES PERMITTING 1-2
1.2.1 BMP Regulatory History 1-3
1.3 BEST MANAGEMENT PRACTICES AND POLLUTION PREVENTION 1-4

2. BEST MANAGEMENT PRACTICES PLAN DEVELOPMENT 2-1
2.1 PURPOSE OF THIS CHAPTER 2-1
2.2 BMP APPLICABILITY 2-2
2.2.1 What Activities and Materials at an Industrial Facility Are Best Addressed by BMP? 2-2
2.2.2 How Do BMPS Work? 2-3
2.2.3 What Are the Types of BMPS? 2-4
2.3 COMPONENTS OF BMP PLANS 2-5
2.3.1 BMP Plan Planning Phase 2-6
2.3.1.1 BMP Committee 2-6
2.3.1.2 BMP Policy Statement 2-11
2.3.1.3 Release Identification and Assessment 2-15
2.3.2 BMP Plan Development Phase 2-23
2.3.2.1 Good Housekeeping 2-24
2.3.2.2 Preventive Maintenance 2-29
2.3.2.3 Inspections 2-35
2.3.2.4 Security 2-40
2.3.2.5 Employee Training 2-43
2.3.2.6 Recordkeeping and Reporting 2-48
2.3.3 BMP Plan Evaluation and Reevaluation Phase 2-54
2.3.3.1 Plan Evaluation 2-54
2.3.3.2 Plan Reevaluation 2-55

3. INDUSTRY-SPECIFIC BEST MANAGEMENT PRACTICES 3-1
3.1 PURPOSE OF THIS CHAPTER 3-1
3.2 INDUSTRY CATEGORY SELECTION 3-2
3.3 METAL FINISHING 3-3
3.3.1 Industry Profile 3-3
3.3.2 Effective BMPs 3-4
3.4 ORGANIC CHEMICALS, PLASTICS, AND SYNTHETIC FIBERS (OCPSF) MANUFACTURING 3-6
3.4.1 Industry Profile 3-6
3.4.2 Effective BMPs 3-7
3.5 TEXTILES MANUFACTURING 3-8
3.5.1 Industry Profile 3-8
3.5.2 Effective BMPs 3-10
3.6 PULP AND PAPER MANUFACTURING 3-11
3.6.1 Industry Profile 3-11

3.6.2 Effective BMPs 3-13
3.7 PESTICIDES FORMULATION 3-14
3.7.1 Industry Profiles 3-14
3.7.2 Effective BMPs 3-15
3.8 PHARMACEUTICALS MANUFACTURING 3-16
3.8.1 Industry Profile 3-16
3.8.2 Effective BMPs 3-17
3.9 PRIMARY METALS MANUFACTURING 3-18
3.9.1 Industry Profile 3-18
3.9.2 Effective BMPs 3-20
3.10 PETROLEUM REFINING 3-21
3.10.1 Industry Profile 3-21
3.10.2 Effective BMPs 3-22
3.11 INORGANIC CHEMICALS MANUFACTURING 3-23
3.11.1 Industry Profile 3-23
3.11.2 Effective BMPs 3-24

4. RESOURCES AVAILABLE FOR DETERMINING BEST MANAGEMENT PRACTICES 4-1
4.1 PURPOSE OF THIS CHAPTER 4-1
4.2 NATIONAL AND INTERNATIONAL RESOURCES 4-1
4.2.1 Pollution Prevention Information Clearinghouse (PPIC) 4-2
4.2.2 International Cleaner Production Information Clearinghouse (ICPIC) 4-5
4.2.3 Waste Reduction Institute for Training and Applications Research, Inc. (WRITAR) 4-6
4.2.4 National Technical Information Service (NTIS) 4-7
4.2.5 Nonpoint Source (NPS) Information Exchange Bulletin Board System (BBS) 4-8
4.2.6 Office of Water Resource Center 4-9
4.3 REGIONAL RESOURCES 4-10
4.3.1 Northeast Multimedia Pollution Prevention (NEMPP) Program 4-10
4.3.2 Waste Reduction Resource Center for the Southeast (WRRC) 4-11
4.3.3 Pacific Northwest Pollution Prevention Research Center (PNPPRC) 4-12
4.3.4 EPA Offices and Libraries 4-13
4.4 STATE, UNIVERSITY, AND OTHER AVAILABLE RESOURCES . . . 4-16
4.4.1 Center for Waste Reduction Technologies (CWRT) 4-16
4.4.2 Solid Waste Information Clearinghouse (SWICH) 4-17
4.4.3 State Resources 4-18
4.4.4 University-Affiliated Resources 4-18

APPENDIX A — BEST MANAGEMENT PRACTICES PLAN DEVELOPMENT CHECKLIST

APPENDIX B — EXAMPLE FORMS AND CHECKLISTS

APPENDIX C — THEORETICAL DECISION-MAKING PROCESS FOR BMP PLAN DEVELOPMENT

APPENDIX D — BIBLIOGRAPHY

LIST OF EXHIBITS

2-1 FACTORS AFFECTING SPECIFIC BMP SELECTION 2-6
2-2 SUGGESTED ELEMENTS OF A BASELINE BMP PLAN 2-7
2-3 BMP COMMITTEE ACTIVITIES AND RESPONSIBILITIES 2-8
2-4 EXAMPLE OF COMMITTEE FORMATION TO EFFECTIVELY MANAGE AN ENVIRONMENTAL PROGRAM 2-9
2-5 EXAMPLE OF THE USE OF A POLICY 2-12
2-6 AN EXAMPLE OF THE EFFECTIVENESS OF USING A RELEASE IDENTIFICATION AND ASSESSMENT APPROACH 2-17
2-7 AN EXAMPLE OF THE SUCCESSFUL IMPLEMENTATION OF A GOOD HOUSEKEEPING PROGRAM 2-26
2-8 CONSIDERATIONS FOR SELECTION OF MITIGATIVE PRACTICES . . . 2-28
2-9 DEMONSTRATION OF THE NEED FOR AN EFFECTIVE PM PROGRAM 2-31
2-10 EXAMPLES OF POOR COORDINATION BETWEEN OPERATIONS AND MAINTENANCE STAFF 2-33
2-11 RELEASES WHICH COULD HAVE BEEN PREVENTED BY EFFECTIVE INSPECTION PROGRAMS 2-36
2-12 POSSIBLE COMPONENTS OF A SECURITY PLAN 2-41
2-13 EXAMPLE OF THE IMPLEMENTATION OF EMPLOYEE TRAINING TO ENSURE THE SUCCESS OF ENVIRONMENTAL CONTROLS 2-44
2-14 EXAMPLE OF AN EFFECTIVE REPORTING PROGRAM DESIGNED TO PREVENT ENVIRONMENTAL RELEASES 2-50
3-1 BMP SELECTION PROCESS 3-1
3-2 SUMMARY OF BMPs UTILIZED IN THE METAL FINISHING INDUSTRY 3-26
3-3 SUMMARY OF BMPs UTILIZED IN THE OCPSF MANUFACTURING INDUSTRY 3-36
3-4 SUMMARY OF BMPs UTILIZED IN THE TEXTILES MANUFACTURING INDUSTRY 3-38
3-5 SUMMARY OF BMPs UTILIZED IN THE PULP AND PAPER MANUFACTURING INDUSTRY 3-43
3-6 SUMMARY OF BMPs UTILIZED IN THE PESTICIDES FORMULATION INDUSTRY 3-48
3-7 SUMMARY OF BMPs UTILIZED IN THE PHARMACEUTICAL MANUFACTURING INDUSTRY 3-50
3-8 SUMMARY OF BMPs UTILIZED IN THE PRIMARY METALS MANUFACTURING INDUSTRY 3-52
3-9 SUMMARY OF BMPs UTILIZED IN THE PETROLEUM REFINING INDUSTRY 3-59
3-10 TYPES OF INORGANIC CHEMICALS 3-24
3-11 SUMMARY OF BMPs UTILIZED IN THE INORGANIC CHEMICALS MANUFACTURING INDUSTRY 3-60
4-1 PPIC CONTACT INFORMATION 4-3
4-2 INSTRUCTIONS FOR PIES USE 4-4
4-3 ICPIC CONTACT INFORMATION 4-6
4-4 WRITAR CONTACT INFORMATION 4-7

4-5 NTIS CONTACT INFORMATION 4-8
4-6 NPS BBS CONTACT INFORMATION 4-9
4-7 OFFICE OF WATER RESOURCE CENTER CONTACT INFORMATION .. 4-10
4-8 NEMPP CONTACT INFORMATION 4-11
4-9 WRRC CONTACT INFORMATION 4-12
4-10 PNPPRC CONTACT INFORMATION 4-13
4-11 EPA REGIONAL POLLUTION PREVENTION CONTACTS 4-14
4-12 EPA LIBRARY CONTACT INFORMATION 4-15
4-13 CWRT CONTACT INFORMATION 4-17
4-14 SWICH CONTACT INFORMATION 4-18
4-15 STATE PROGRAM INFORMATION 4-19
4-16 UNIVERSITY AFFILIATED RESOURCES 4-28

1. INTRODUCTION TO BEST MANAGEMENT PRACTICES

Best management practices (BMPs) are recognized as an important part of the National Pollutant Discharge Elimination System (NPDES) permitting process to prevent the release of toxic and hazardous chemicals. Over the years, as BMPs for many different types of facilities have been developed, case studies have demonstrated not only the success but the flexibility of the BMP approach in controlling releases of pollutants to receiving waters. More recently, pollution prevention practices have become part of the NPDES program, working in conjunction with BMPs to reduce potential pollutant releases. Pollution prevention methods have been shown to reduce costs as well as pollution risks through source reduction and recycling/reuse techniques.

The Federal Water Pollution Control Act of 1972 established the objectives of restoring and maintaining the chemical, physical, and biological integrity of the Nation's waters. These objectives remained unchanged in the 1977, 1982, and 1987 amendments, commonly referred to as the Clean Water Act (CWA). To achieve these objectives, the CWA sets forth a series of goals, including attaining fishable and swimmable designations by 1983 and eliminating the discharge of pollutants into navigable waters by 1985. As part of the CWA strategy to eliminate discharges of pollutants to receiving waters, NPDES permit limitations have become more stringent. The Environmental Protection Agency (EPA) recognizes that industrial and municipal facilities subject to the NPDES program may need to undertake additional measures to meet these permit limitations, as well as the goals of the CWA. EPA believes that such measures can be technologically and economically achievable through the development of formalized plans that contain BMPs and pollution prevention practices.

1.1 PURPOSE OF THIS MANUAL

The purpose of this manual is to provide guidance to NPDES permittees in the development of BMPs for their facilities. The manual may also be useful to NPDES permit writers and inspectors charged with evaluating the adequacy of BMP plans. In particular, the manual promotes the integration of pollution prevention concepts and practices in BMP plans. This manual has four major goals: (1) to provide a general understanding of the requirements of the CWA pertaining to BMPs and show the relationship between BMPs and pollution prevention practices; (2) to provide a starting point for developing and implementing an effective BMP plan that integrates facility-specific and general BMPs and pollution prevention practices; (3) to provide specific examples of effective BMPs and pollution prevention practices to aid facilities endeavoring to develop their own BMPs; and finally (4) to identify sources which a facility may consult when developing BMPs and pollution prevention practices.

This manual includes four chapters intended to achieve the above goals. Chapter 1 provides an introduction to the NPDES program and the regulatory context for BMPs. The relationship between BMPs and other pollution prevention requirements such as the Pollutant Prevention Act is also discussed. Chapter 2 discusses the suggested components of a BMP plan, including committee formation, policy derivation, release identification and assessment, good housekeeping, preventive maintenance, inspections, security, employee training, and recordkeeping and reporting. Each component is defined and described in terms of what the component is, how the component functions, methods to create the component, and what to do/what not to do. Additionally, the usefulness of each component is illustrated by an example, thereby promoting the development of an effective BMP plan. Chapter 3 sets forth process-specific BMPs for the metal plating and finishing, pesticides, textiles, pulp and paper, organic chemicals, pharmaceuticals, primary metals manufacturing and forming, inorganic chemicals, and petroleum refining industries. Successful and demonstrated BMPs are discussed in this chapter in terms of benefits to water, benefits to other media, and other incentives. Data sources are also cited to enable readers to consult the referenced document. Finally, a discussion of available resources at the international, national, regional, and State levels is presented in Chapter 4. Programs are summarized in terms of the general resources available and limitations in the scope of assistance. Specific information is provided to enable readers to contact programs directly and obtain necessary information.

1.2 BACKGROUND OF NPDES PERMITTING

The principal mechanism for reducing the discharge of pollutants from point sources is through implementation of the NPDES program, established by Section 402 of the CWA. All facilities with point source discharges must apply for and obtain a NPDES permit. NPDES-authorized States are tasked with issuing permits. Where State NPDES authorization has not yet occurred, EPA Regions issue NPDES permits.

Four minimum elements are typically included in each permit issued: (1) effluent discharge limitations; (2) monitoring and reporting requirements; (3) standard conditions; and (4) special conditions. The numeric effluent discharge limits contained in a NPDES permit are based on the most stringent value among technology-based effluent guidelines limitations, water quality-based limitations, and limitations derived on a case-by-case basis. Permits also contain standard conditions that prescribe primarily administrative and legal requirements to which all facilities are subject. Finally, permits may contain any supplemental controls, referred to as special conditions, that may be needed in order to ensure that the regulations driving the NPDES program and, ultimately, the goals of the CWA are met. Best management practices are one such type of supplemental control.

1.2.1 BMP Regulatory History

Section 304(e) of the CWA authorized the EPA Administrator to publish regulations to control discharges of significant amounts of toxic pollutants listed under Section 307 or hazardous substances listed under Section 311 from industrial activities that the Administrator determines are associated with or ancillary to industrial manufacturing or treatment processes. As defined by the CWA, the discharges to be controlled by BMPs are plant site runoff, spillage or leaks, sludge or waste disposal, and drainage from raw material storage.

On September 1, 1978, EPA proposed regulations (43 FR 39282) addressing the use of procedures and practices to control discharges from activities associated with or ancillary to industrial manufacturing or treatment processes. The proposed rule indicated how BMPs would be imposed in NPDES permits to prevent the release of toxic and hazardous pollutants to surface waters. The regulations (40 CFR Part 125, Subpart K, Criteria and Standards for Best Management Practices Authorized Under Section 304(e) of the CWA) were proposed on August 21, 1978, in the NPDES regulations (43 FR 37078). While this Subpart never became effective, it remains in the Code of Federal Regulations and can be used as guidance by permit writers.

Although 40 CFR Part 125, Subpart K was not finalized, EPA and States continue to incorporate BMPs into permits based on the authority contained in Section 304(e) of the CWA and the regulations set forth in 40 CFR 122.44(k). While Section 304(e) of the CWA restricts the application of BMPs to ancillary sources and certain chemicals, the regulations contained in 40 CFR 122.44(k) authorize the use of BMPs to abate the discharge of pollutants when (1) they are developed in accordance with Section 304(e) of the CWA, (2) numeric limitations are infeasible, or (3) the practices are necessary to achieve limitations/standards or meet the intent of the CWA. Thus, permit writers are afforded considerable latitude in employing BMPs as pollution control mechanisms.

1.3 BEST MANAGEMENT PRACTICES AND POLLUTION PREVENTION

Best management practices are inherently pollution prevention practices. Traditionally, BMPs have focused on good housekeeping measures and good management techniques intending to avoid contact between pollutants and water media as a result of leaks, spills, and improper waste disposal. However, based on the authority granted under the regulations, BMPs may include the universe of pollution prevention encompassing production modifications, operational changes, materials substitution, materials and water conservation, and other such measures.

EPA endorses pollution prevention as one of the best means of pollution control. In 1990, the Pollution Prevention Act was enacted and set forth a national policy that:

> "... pollution should be prevented or reduced at the source whenever feasible; pollution that cannot be prevented should be recycled in an environmentally safe manner, whenever feasible; pollution that cannot be prevented or recycled should be treated in an environmentally safe manner whenever feasible; and disposal or other release into the environment should be employed only as a last resort and should be conducted in an environmentally safe manner."

EPA recognizes that significant opportunities exist for industry to reduce or prevent pollution through cost-effective changes in production, operation, and raw materials use. In addition, such changes may offer industry substantial savings in reduced raw materials, pollution control, and liability costs, as well as protect the environment and reduce health and safety risks to workers. Where pollution prevention practices can be both environmentally beneficial and economically feasible, EPA finds their implementation to be prudent.

EPA believes that the intent of pollution prevention practices and BMPs are similar and that they can be concurrently developed in a technologically sound and cost-effective manner. Thus, although this manual primarily focuses on best management practices, which pertain to the NPDES program, the reader may be compelled to also consider pollution prevention for all media in order to maximize the benefits achieved.

2. BEST MANAGEMENT PRACTICES PLAN DEVELOPMENT

Many facilities currently implement successful measures to reduce and control environmental releases of all types of pollutants. These measures have been successfully implemented both formally as part of best management practice (BMP) plans and informally as part of unwritten standard operating procedures. In the context of the NPDES permit program, permittees are required to develop BMP plans to address specific areas of concern. The BMP plan developed by the permittee becomes an enforceable condition of the permit. The Environmental Protection Agency (EPA) believes that to ensure the continuing and greater successes of these programs, pollution prevention measures should be incorporated into a written company-wide plan.

2.1 PURPOSE OF THIS CHAPTER

This chapter provides the reader with the information needed to develop and implement a BMP plan. The chapter begins with a discussion of applicability of BMP plans to industrial facilities. The remainder of the chapter provides a detailed discussion of each of the recommended components of a BMP plan, set forth in the following format:

- ***What is the component?*** A description is given including definitions, applicability, and general limitations.
- ***How does the component function?*** The text discusses how the component interacts with other components, considerations when developing the component, and an outline of the steps involved in the development process.
- ***How is the component created/developed?*** An explanation of the detailed steps to take in developing a program around the element is provided.
- ***What to do/what not to do.*** Guidance has been offered in terms of helpful hints as well as potential problems to avoid.

To increase the usefulness of the information, examples of actual BMPs accompany the text. Additionally, three appendices supplement this Chapter. Considerations when planning and developing the BMP plan are set forth in a checklist format in Appendix A. Example forms and checklists that may be useful to facilities in the development and implementation of BMP plan activities are provided in Appendix B. Appendix C contains an example BMP plan and the decision-making process used during its development.

2.2 BMP APPLICABILITY

BMPs are developed as part of the National Pollutant Discharge Elimination System (NPDES) permitting requirements to control releases of harmful substances. BMPs may apply to an entire site or be appropriate for discrete areas of an industrial facility. Many of the same environmental controls promoted as part of a BMP plan may currently be used by industry in storm water pollution prevention plans, spill prevention control and countermeasure (SPCC) plans, Occupational Safety and Health Administration (OSHA) safety programs, fire protection programs, insurance policy requirements, or standard operating procedures. Additionally, where facilities have developed pollution prevention programs, controls such as source reduction and recycling/reuse may be similar to those promoted as part of a BMP plan. The following basic questions can be used to establish the scope of BMP plans: (1) What activities and materials at an industrial facility are best addressed by BMPs? (2) How do BMPs work? and (3) What are the types of BMPs?

2.2.1 What Activities and Materials at an Industrial Facility Are Best Addressed by BMPs?

Traditionally, BMP activities have focussed on activities associated with or ancillary to industrial manufacturing or treatment processes. These have been identified in Section 304(e) of the Clean Water Act as "plant site runoff, spillage or leaks, sludge or waste disposal, and drainage from raw material storage which the Administrator determines are associated with or ancillary to the industrial manufacturing or treatment process." These activities have historically been found to be amenable to control by BMPs. Some examples include the following:

- ***Material storage areas*** for toxic, hazardous, and other chemicals including raw materials, intermediates, final products, or byproducts. Storage areas may be piles of materials or containerized substances. Typical storage containers could include liquid storage vessels ranging in size from large tanks to 55-gallon drums; dry storage in bags, bins, silos and boxes; and gas storage in tanks and vessels. The storage areas can be open to the environment, partially enclosed, or fully contained.

- ***Loading and unloading operations*** involving the transfer of materials to and from trucks or rail cars, including in-plant transfers. These operations include pumping of liquids or gases from truck or rail car to a storage facility or vice versa, pneumatic transfer of dry chemicals during vehicle loading or unloading, transfer by mechanical conveyor systems, and transfer of bags, boxes, drums, or other containers from vehicles by fork-lift, hand, or other materials handling methods.

- ***Facility runoff*** generated principally from rainfall on a plant site. Runoff can become contaminated with harmful substances when it comes in contact with material storage areas, loading and unloading areas, in-plant transfers areas, and sludge and other waste storage/disposal sites. Fallout, resulting from plant air emissions that settle on the plant site, may also contribute to contaminated runoff. In addition to BMPs, facility runoff from industrial sites may also be directly regulated under the NPDES storm water permitting program.

- ***Sludge and waste storage and disposal areas*** including landfills, pits, ponds, lagoons, and deep-well injection sites. Depending on the construction and operation of these sites, there may be a potential for leaching of toxic pollutants or hazardous substances to groundwater, which can eventually reach surface waters. In addition, liquids may overflow to surface waters from these disposal operations.

With the increasing in awareness of pollution prevention opportunities as well as the increase in legislation and regulatory policies directing efforts towards pollution prevention, much of the traditional focus of BMP activities is being redirected from ancillary activities to industrial manufacturing processes. This redirection is resulting in the integrated application of traditional BMPs and pollution prevention practices into cohesive and encompassing plans that cover all aspects of industrial facilities.

2.2.2 How Do BMPs Work?

- ***BMPs are practices or procedures***. They include methods to prevent toxic and hazardous substances from reaching receiving waters. They are most effective when organized into a comprehensive facility BMP plan.

- ***BMPs are qualitative***. They are designed to address the quality of a facility's practices, and may ultimately affect the ability of the facility to meet environmental control standards.

- ***BMPs are flexible.*** Many different practices can be used to achieve similar environmentally protective results. With facility-specific considerations as the major consideration in selecting appropriate BMPs, this flexibility allows a facility to tailor a BMP plan to meet its needs using the capabilities and resources available.

2.2.3 What Are the Types of BMPs?

BMPs may be divided into general BMPs, applicable to a wide range of industrial operations, and facility-specific (or process-specific) BMPs, tailored to the requirements of an individual site. General BMPs are widely practiced measures that are independent of chemical compound, source of pollutant, or industrial category. General BMPs are also referred to as baseline practices, and are typically low in cost and easily implemented. General BMPs are practiced to some extent at almost all facilities. Common general BMPs include good housekeeping, preventive maintenance, inspections, security, employee training, and recordkeeping and reporting.

Facility-specific BMPs are measures used to control releases associated with individually identified toxic and hazardous substances and/or one or more particular ancillary source. Facility-specific BMPs are often developed when a facility notes a history of problem releases of toxic or hazardous chemicals, or when facility personnel believe that actual or potential pollutant discharge problems should be addressed. Facility-specific BMPs may include many different practices such as source reduction and on-site recycle/reuse.

Facility-specific BMPs will vary from site to site depending upon site characteristics, industrial processes, and pollutants. For example, a site-specific BMP in the form of area dikes may be adopted due to the location of the facility: facilities in flat areas or on slopes are likely to utilize dikes to control spills whereas there may be no need for dikes for facilities located in basins. Additionally, plants handling and storing large amounts of liquid chemicals would be more likely to utilize dikes than facilities storing and using dry chemicals. Facilities experiencing erosion and sediment control problems may consider establishing vegetative buffer strips or indigenous ground cover for purposes of soil stabilization and infiltration of runoff. Facility sites located adjacent to other industrial areas may consider runon controls to prevent extraneous spills and contaminated runon from entering the facility site. Other site-specific considerations, such as endangered species, may motivate facilities to store materials in an alternate location so as to prevent exposure.

Processes also drive the determination of appropriate specific BMPs. Materials handling procedures that expose employees to toxic chemicals (e.g., hand drawing) may prompt the consideration of procedures that reduce the potential for exposure (e.g., automated pneumatic pumping in enclosed conduits). Some examples of process-specific BMPs include the following:

- Using splash plates designed to prevent spills at a metal finishing facility
- Installing solvent recovery equipment to control benzene releases at a petroleum refinery
- Purchasing solvents in reusable containers rather than 55-gallon drums to ease storage concerns and reduce wastewater resulting from requirements to triple rinse drums.

Pollutant characteristics such as volatility and toxicity also affect BMP selection. More and more, harmful chemicals are being considered for replacement with less toxic alternatives, or for elimination from the process. Ozone layer-depleting solvents which are used for cleaning at many facilities are being replaced with detergent-based cleaning agents. Additionally, facilities using materials with toxic properties have been inspired to take more proactive control measures (e.g., double walled containment).

The choice of facility-specific BMPs can be affected by a number of factors such as those discussed in Exhibit 2-1.

2.3 COMPONENTS OF BMP PLANS

Suggested components of BMP plans are defined and described in this section. The suggested elements of a good BMP plan can be separated into three phases: planning, development and implementation, and evaluation/reevaluation. Generally, the planning phase, discussed in Section 2.3.1, includes demonstrating management support for the BMP plan and identifying and evaluating areas of the facility to be addressed by BMPs. The goal of plan development should be to ensure that its implementation will prevent or minimize the generation and the potential for release of pollutants from the facility to the waters of the U.S. The development phase consists of determining, developing, and implementing general and facility-specific BMPs and is described in Section 2.3.2. The evaluation/reevaluation phase described in Section 2.3.3 consists of an assessment of the components of a BMP plan and reevaluation of plan components periodically, or as a result of factors such as environmental releases and/or changes at the facility. Suggested elements of a BMP plan are provided in Exhibit 2-2.

EXHIBIT 2-1: FACTORS AFFECTING SPECIFIC BMP SELECTION

- **Chemical nature**: The need to control materials based on toxicity and fate and transport.
- **Proximity to waterbodies**: The need to control liquid spills prior to their release to media such as water from which materials may not later be separated.
- **Receiving waters**: The need to protect sensitive receiving waters which are more severely impacted by releases of toxic or hazardous materials. The need to protect the water uses including recreational waters, drinking water supplies, and fragile aquatic and biota communities.
- **Proximity to populace**: The need to control hazardous materials with potential to be released near populated areas.
- **Climate**: The need to prevent volatilization and ignitability in warmer climates. The need to reduce wear on moving parts in freezing climates. The need to avoid spills in climates and under circumstances where mitigation cannot occur.
- **Age of the facility/equipment**: The need to prevent releases caused by older equipment with greater capacity for failure. The need to address obsolete and outdated instruments and processes which are not environmentally protective.
- **Process complexity**: The need to address problems of materials incompatibility.
- **Engineering design**: The need to address design flaws and deficiencies.
- **Employee safety**: The need to prevent unnecessary exposure between employee and chemicals.
- **Environmental release record**: The need to control releases from specific areas demonstrating previous problems.

2.3.1 BMP Plan Planning Phase

In the planning phase, a facility must decide who will take the responsibility for establishing and carrying out the BMP plan. The plan should be initiated with clear support and input from facility management and employees. The facility must also identify and evaluate areas of the facility that, because of the substances involved and their management, will be addressed in the BMP plan. Each of these elements is discussed in detail on the following pages.

2.3.1.1 BMP Committee

What is a BMP Committee?

A BMP committee is comprised of interested staff within the facility's organization. The committee will represent the company's interests in all phases of BMP plan development, implementation, oversight, and plan evaluation.

It should be noted that a BMP committee may function similarly to other committees that might exist at an industrial facility (e.g., pollution prevention committee) and may include the same employees.

How Does the BMP Committee Function?

The BMP committee is developed to assist a facility in managing all aspects of the BMP plan. The committee functions to conduct activities and shoulder the responsibilities of all elements discussed in Exhibit 2-3.

EXHIBIT 2-2: SUGGESTED ELEMENTS OF A BASELINE BMP PLAN

Planning Phase Considerations

1. BMP committee
2. BMP policy statement
3. Release identification and assessment

Development Phase Considerations

4. Good housekeeping
5. Preventive maintenance
6. Inspections
7. Security
8. Employee training
9. Recordkeeping and reporting

Evaluation and Reevaluation Phase Considerations

10. Evaluate plan implementation benefits
11. Periodically or as needed, repeat steps 1-9

To be most effective, the committee must perform tasks efficiently and smoothly. In large part, the personnel selected to act as committee members will determine the committee's success. Some of the considerations for personnel selection include the following:

- A lead committee member must be determined

- Committee members must include persons knowledgeable of the plant areas involved (e.g., process areas, tank farms) and utilization of chemicals and generation of pollutants (e.g., solvents, products, chemical reactants) at the facility

- Committee members should have the authority to make decisions effecting BMP plan development and implementation

- The size of the committee must be appropriate to the facility.

- The committee must represent affected areas of the plant and employees.

An example of the effectiveness of the formation of a committee is provided in Exhibit 2-4.

EXHIBIT 2-3: BMP COMMITTEE ACTIVITIES & RESPONSIBILITIES

- Develop the scope of the BMP plan
- Make recommendations to management in support of company BMP policy
- Review any existing accidental spill control plans to evaluate existing BMPs
- Identify toxic and hazardous substances
- Identify areas with potential for release to the environment
- Conduct assessments to prioritize substances and areas of concern
- Determine and select appropriate BMPs
- Set forth standard operating procedures for implementation of BMPs
- Oversee the implementation of the BMPs
- Establish procedures for recordkeeping and reporting
- Coordinate facility environmental release response, cleanup, and regulatory agency notification procedures
- Establish BMP training for plant and contractor personnel
- Evaluate the effectiveness of the BMP plan in preventing and mitigating releases of pollutants
- Periodically review the BMP plan to evaluate the need to update and/or modify the BMP plan.

How Is a BMP Committee Developed?

The BMP committee is responsible for developing the BMP plan and assisting the facility management in its implementation, periodic evaluation, and updating. While the BMP committee is responsible for developing the plan and overseeing its implementation, all activities need not be limited to committee members. Rather, appropriate company personnel who are knowledgeable in the areas of concern can carry out certain activities associated with BMP plan development. With this in mind, the selection of the committee members can be limited to a select set of individuals, while the resources of interested and knowledgeable employees can still be utilized.

In order to ensure a properly run organization, one person should be designated as the lead committee member. Thus, the first step in developing a BMP committee is to determine the appropriate committee chairperson. The determination of a single leader will assist in the smooth conduct of meetings and the designation of tasks, and will aid in the decision-making process. Generally, the chairperson should be highly motivated to develop and implement the BMP plan, familiar with all committee members and their areas of expertise, and experienced in managing tasks

EXHIBIT 2-4: EXAMPLE OF COMMITTEE FORMATION TO EFFECTIVELY MANAGE AN ENVIRONMENTAL PROGRAM

The 3M company, a manufacturer of diverse products such as coated abrasives, pressure sensitive tape, photographic film, electrical insulation materials, and reposition notes, has developed a corporate philosophy that Pollution Prevention Pays (referred to as the 3P program). As part of the 3P program, 3M has created a 3P Coordinating Committee which includes employee representatives from the engineering manufacturing, laboratory, and corporate environmental sectors.

The 3P Coordinating Committee provides support and coordination for nationwide teams establishing 3P programs. These 3P Teams are organized by employees that have identified pollution problems and recognize potential solutions.

The 3P Coordinating Committee and the 3P Teams have been instrumental in source reduction of hydrocarbons, odor, water, dissolved solids, sulfur, zinc, alcohol, and incinerated scrap. In the first year of the 3P programs, air pollutants have been reduced by 123,000 tons, water pollutants by 16,400 tons, wastewater by 1,600 million gallons, and solid wastes pollutants by 409,000 tons. This has resulted in savings of more than $500 million.

Adapted from: T. Zeal, "Case Study: How 3M Makes Pollution Prevention Pay Big Dividends," *Pollution Prevention Review*, Winter 1990-91.

of this magnitude. The chairperson will be responsible for ensuring that all tasks are assigned to appropriate personnel, keeping facility management and employees informed, and cohesively developing the BMP plan. Potential candidates for this role are plant managers, environmental coordinators, or other distinctly knowledgeable technical and management personnel.

The next step is to select the appropriate personnel to comprise the committee. Personnel selected should represent all affected facility areas. Members might also be selected based on their areas of expertise (e.g., industrial processes). Personnel might be selected who have a full understanding of the manufacture processes from raw materials to final products, as well as of the recycling, treatment, and disposal of wastes. Possible candidates include foremen in manufacturing, production, or waste treatment and disposal; maintenance engineers; environmental and safety coordinators; and materials storage and transfer managers. Not only must committee members understand the activities conducted throughout the entire facility, members of the BMP committee must also include individuals who are in the decision-making positions within the company structure. Some committee members must represent company management and have the authority to implement measures adopted by the committee.

While the BMP committee should reflect the lines of authority within the company, it should also be sensitive to general employee interests. It is crucial to ensure that employees are aware of and in support of the BMP plan and the responsible committee, as it is primarily the employees who will implement the changes resulting from committee decisions. Forming a committee comprised solely of upper level management and administrative personnel would exclude general personnel whose input is critical for the development and implementation of the plan. Selecting employee-chosen representatives, such as union stewards, may be an appropriate means to ensure employee involvement.

The size of a BMP committee should reflect the size and complexity of the facility, as well as the quantity and toxicity of the materials at the facility. The committee must be small enough to communicate in a open and interactive manner, yet large enough to allow for input from all necessary parties.

Where needed, committee members should call upon the expertise of others through the establishment of project-specific task forces. For example, personnel involved in research and development may be asked to research the effectiveness of product substitution and process changes that are being considered as part of BMP plan development. This method of calling upon specialists, when the need arises, should allow the committee to remain a manageable size. Generally, the size selection process outlined below presents a good rule of thumb:

- For small facilities, a single committee member is acceptable as long as that person has the requisite expertise and authority
- For larger facilities, selection of six to eight people as permanent members of the committee should be ideal.

BMP Committee - What to Do

- Develop a roster of BMP committee members which includes area of specialization and projected responsibilities. This list helps identify any holes in the planned BMP development activities and any missing expertise.

- Include a list of alternate BMP committee members where transfers are expected to occur during the life of the BMP plan.

- Post BMP committee member names and including them in the plan to allow any interested parties the opportunity to contact BMP committee members.

- While developing and updating the BMP plan, include input from interested employees not on the committee. Employee input sessions and suggestions boxes can be used to meet this goal.

- Extend technical reviews to personnel not on the BMP committee, where specialized expertise is necessary or where interest is expressed.

- Follow up with all responsible parties on a periodic basis to ensure they are aware of their BMP-related responsibilities.

- Encourage BMP committee members to spend time on-the-line in order to communicate with other potentially interested parties.

- Set schedules with milestone dates for the performance of important activities. This avoids possible procrastination and allows the BMP plan development to remain on schedule.

BMP Committee - What Not to Do

- The committee should enable, not impede, the decision-making process for preventing or mitigating spills or otherwise responding to events addressed by the BMP plan.

- Remember that personnel contributing to the design of a BMP need not be member of the BMP committee. This is of particular importance where a technical specialist or manager simply would not have the time to contribute on a regular basis.

2.3.1.2 BMP Policy Statement

What Is a BMP Policy Statement?

A BMP policy statement describes the objectives of the BMP program in clear, concise language and establishes the company policies related to BMPs. Exhibit 2-5 provides examples of the successful use of a policy.

EXHIBIT 2-5: EXAMPLE OF THE USE OF A POLICY

Dow Chemical has observed a significant impact as a result of their company's environmental policy. As part of their "Environmental Policy and Guidelines," Dow has set forth a hierarchy similar to that developed as part of the Pollution Prevention Act of 1990. Dow's policy sets forth preferences to handle materials by reducing pollutants at the source, followed by recycle and use of materials whenever possible. Where disposal is necessary, Dow has specified that incineration be considered first, followed by land disposal on Dow-owned property, and finally land disposal on property not owned by Dow.

Dow's decision to follow its disposal hierarchy was based in large part on the liability of disposal. Dow reasoned that incineration was the most appropriate disposal method since it resulted in the pollutants in the ash materials being in elemental form. In many cases, the company has identified opportunities for recycle of materials found in the incinerator ash. Dow also imposed a $215 per drum surcharge for hazardous wastes going to a landfill to provide incentives for finding alternatives to landfilling. Dow also believes that they can better exercise control of onsite disposal, thus influencing the preferences for onsite rather than offsite disposal.

Dow's policy has resulted in an impact on the environmental releases. For example, at the Dow Pittsburg, California facility, wastewater discharges have been reduced by 95 percent over the past 10 years. Additionally, the approximately 10.2 million pounds of chlorinated organics wastes which are generated are either incinerated or recycled.

Adapted from: D. Sarokin, W. Muir, C. Miller, S. Sperber, *Cutting Chemical Wastes: What 29 organic Chemical Plants are Doing to Reduce Hazardous Wastes,* INFORM, Inc., New York, New York, 1985.

How Does a BMP Policy Statement Function?

The policy statement provides two major functions: (1) it demonstrates and reinforces management's support of the BMP plan; and (2) it describes the intent and goals of the BMP plan. It is very important that the BMP policy represent both the company's goals and general employee concerns. Several steps to take in developing an effective BMP policy statement include the following:

- Determine the appropriate author
- Develop tone and content that are positive, but that establish realistic and achievable goals
- Distribute the policy statement effectively.

How Is a BMP Policy Statement Created?

The first step in creating a BMP policy statement is determining the appropriate author. To indicate management's commitment, the policy statement should be signed by a responsible corporate officer. A responsible corporate officer can be the president, vice president, or the principal manager of manufacturing, production, or operations. Generally, the policy statement author should be a person who performs policy- or decision-making functions for the corporation/facility.

The next step in developing the BMP policy statement is to craft the specific language. The policy statement may include references to the company's commitment to being a good environmental citizen, expected improvements in plant safety, and potential cost savings. Regardless of personal style, in all cases the policy should: (1) indicate the company's support of BMPs to improve overall facility management and (2) introduce the intent of the BMP plan.

The length and level of detail of the policy statement will vary depending on the writer's personal style. The following variations may be included in a BMP policy statement:

- An outline of steps that will be taken
- A discussion of the time frames for development and implementation
- An indication of the areas and pollutants of focus
- A projection of the end result of the BMP plan
- Create enthusiasm and support for the BMP plan by all employees.

The tone of the BMP policy statement is also important. The projected positive impacts of BMP implementation should be discussed in general terms. If specific goals are outlined, the level of information and the expectations presented should be reasonable, to avoid overwhelming the reader. Ultimately, the policy should provide an upbeat message of the improved working environment that will result from BMP implementation. Since gaining employee support is so important, it may be appropriate to solicit employee concerns prior to the development of the BMP policy. These concerns can be highlighted as areas which will be evaluated during BMP plan development.

Finally, to ensure that all employees are aware of the impending BMP plan, the policy statement should be printed on company letterhead (for an official appearance) and distributed to all employees. Complete distribution can be best ensured if the statement is both delivered to each employee and posted in common areas.

BMP Policy Statement - What to Do

- Utilize meetings and open sessions to solicit employee participation in the development of the BMP policy.
- Demonstrate that employee ideas are welcome by immediate follow-up on suggestions. Discuss possible implementation opportunities or reasons that implementation is feasible.
- Keep the statement clear and concise.
- Post the BMP policy statement in key locations where employees congregate so that employees will discuss it.
- Use the policy statement to promote an emblem/motto that represents the BMP plan and its benefits.
- Emboss the objectives of the BMP policy on a plaque.

BMP Policy Statement - What Not to Do

- Do not include details of the BMP plan in the policy statement.
- The BMP policy statement should not be issued solely by the BMP committee. It should be issued by the company.

2.3.1.3 Release Identification and Assessment

What Is Release Identification and Assessment?

Release identification is the systematic cataloging of areas at a facility with ongoing or potential releases to the environment. A release assessment is used to determine the impacts on human health and the environment of any on-going or potential releases identified. The identification and assessment process involves the evaluation of both current discharges and potential discharges.

The release identification and assessment process can provide a focus for the range of BMPs being considered on those activities and areas of a facility where the risks (considering the potential for release and the hazard posed) are the greatest. In some cases, the assessment may be performed based on experience and knowledge of the substances and circumstances involved. In other cases, more detailed analyses may be necessary to provide the correct focus, and release assessments may then rely on some of the techniques of risk assessment (e.g., pathway analysis, toxicity, relative risk). Understanding the dangers of releases involves both an understanding of the hazards each potential pollutant poses to human health and the environment, as well as the probability of release due to the facility's methods of storage, handling, and/or transportation.

Some facilities may identify a number of situations or circumstances representing actual or potential hazards that should all be addressed in some detail through the BMP plan. However, in some instances prioritizing potential hazards is the most sensible and cost effective approach. The following example illustrates the need for BMP prioritization: ACME Concrete is a concrete and supply facility with a designated area used to house maintenance vehicles and materials as well as to stockpile construction materials and equipment. Among other things, this facility contains a large stockpile of building sand used to prepare concrete, a vehicle maintenance area where oil is drained from company vehicles, and a shed where drums of solvents used in cleaning operations are stored. Although each of the three materials mentioned at the site (sand, used oil, and solvents) can cause environmental or health damage unless they are controlled, it would not be feasible or reasonable to control losses of small amounts of clean building sand with the same careful attention given to the release of toxic solvents. As this simplistic example shows, priorities for BMPs should reflect a basic understanding of the loss potential and hazards posed by these potential losses. A prudent manager of the ACME Concrete maintenance yard could first limit the potential for escape of solvents through careful training and periodic preventative maintenance and inspection of drums and storage facilities, then prevent runoff of used oil to surface waters or groundwater by collecting and recycling used oil, and finally control major losses of sand through constructions of filter fences or sediment ponds.

How Does Release Potential Identification and Assessment Function?

Identifying and assessing the risk of pollutant releases for purposes of a BMP plan can best be accomplished in accordance with a five-step procedure:

- Reviewing existing materials and plans, as a source of information, to ensure consistency, and to eliminate duplication
- Characterizing actual and potential pollutant sources that might be subject to release
- Evaluating potential pollutants based on the hazards they present to human health and the environment
- Identifying pathways through which pollutants identified at the site might reach environmental and human receptors
- Prioritizing potential releases.

Once established, these priorities may be used in developing a BMP plan that places the greatest emphasis on the sources with the greatest overall risk to human health and the environment, considering the likelihood of release and the potential hazards if a release should occur, while still implementing low cost BMPs that might contribute to safety or other worker driven needs. An example of the effectiveness of this type of assessment is provided in Exhibit 2-6.

An example of a release identification and assessment worksheet is provided in Appendix B. A completed version is also provided to demonstrate how this worksheet can assist facilities in data compilation.

EXHIBIT 2-6: AN EXAMPLE OF THE EFFECTIVENESS OF USING A RELEASE IDENTIFICATION AND ASSESSMENT APPROACH

Borden Chemical company conducted an evaluation of their Fremont, California, facility consistent with the approach for a release potential identification and assessment. Initially, Borden conducted extensive monitoring to determine the sources of organic loadings. Then, knowledgeable plant personnel conducted a plant inspection to identify and assess sources of organic loadings. The comprehensive inspection involved the evaluation of the entire site from initial materials arrival, through production, to final product shipment. The Borden staff were considerate of both actual and potential sources. Based on the information from the monitoring program and the inspection, the plant management prioritized three areas in which modifications were needed: filter rinse operations, reactor vessel rinses, and employee practices.

Based on this assessment, Borden implemented a system which involves process changes, employee training, and a continuing monitoring program. Ultimately, the amount of organic materials discharged was reduced by 93 percent.

Adapted from: D. Sarokin,W. Muir,C. Miller,S. Sperber,*Cutting Chemical Wastes: What 29 Organic Chemical Plants are Doing to Reduce Hazardous Waste*, Inform, Inc., New York, New York, 1985.

How Is a Release Potential Identification and Assessment Performed?

The first step in the conduct of a release identification and assessment involves the review of existing materials and plans to gather needed information. Many industrial facilities are already subject to regulatory requirements to collect and provide information that may be useful in the identification and assessment of releases. In some cases, these plans may have been developed by persons in plant safety or process engineering who do not normally consider themselves part of the environmental staff. In particular, the following plans should be identified and reviewed:

- ***Preparedness, prevention, and contingency plans (see 40 CFR Parts 264 and 265)*** require the identification of hazardous wastes handled at a facility.
- ***Spill control and countermeasures (SPCC) plans (see 40 CFR Part 112)*** require the prediction of direction, rate of flow, and total quantity of oil that could be discharged.

- ***Storm water pollution prevention plans (see 40 CFR 122.44)*** require the identification of potential pollutant sources which may reasonably be expected to affect the quality of storm water discharges.

- ***Toxic organic management plans (see 40 CFR Parts 413, 433, and 469)*** may require the identification of toxic organic compounds.

- ***Occupational Safety and Health Administration (OSHA) emergency action plans (see 29 CFR Part 1910)*** requires the development of a list of major workplace fire and emergency hazards.

Other sources of information that might be pertinent to the release identification and assessment process include the facility's NPDES permit application and, where applicable, information collected for SARA Title III, Section 313 Form R . SARA requires facilities with certain chemicals to annually submit toxic release data annually as part of community right-to-know requirements.

The second step of conducting a release identification and assessment is to characterize current and potential pollutant sources. This step may be conducted through assembling a description of facility operations and chemical usage and then verifying information through inspections. This process allows facility personnel to confirm the accuracy of information on hand (e.g., the amount of chemicals used in a specific location) while also tracking changes that might have evolved over time (e.g., changing the staging of lubricants in a particular part of the plant).

Generally, the preparation of a site map or maps covering the entire facility is very useful in this evaluation. Maps should cover the entire property and illustrate plant features including material storage areas for raw materials, by-products, and products; loading and unloading areas; manufacturing areas; and waste/wastewater management areas. The map should also indicate site topography including facility drainage patterns. Any existing structural control measures already used to reduce pollutant releases should be highlighted, and conveyance mechanisms or pathways to surface water bodies should be noted. The facility site map should also indicate property boundaries, buildings, and operation or process areas. Any neighboring properties that have potential sources of contaminants that might migrate onto the facility (because of drainage patterns) should also be noted on the map.

Following preparation of a site map, a materials inventory should be prepared. Generally, purchasing records should be helpful in determining the raw materials which are part of the inventory. However, the products manufactured and the byproducts resulting during the manufacturing process should also be considered.

The materials inventory should include descriptions of the amounts of pollutants released or with the potential to be released based on methods of storage or onsite disposal, loading and access methods, and management and control practices (including structural measures or treatment). The inventory should refer to the location of the material keyed to the site map. Materials inventories will vary with the size and the complexity of the facility. It may be helpful to conduct separate inventories for different areas (e.g., manufacturing areas 1, 2, and 3; water drainage areas 1 and 2).

The site map and materials inventory developed to this point have relied solely on plant records. The next part of this process requires a field evaluation/inspection that verifies the facts compiled to this point, and determines the reasons for any discrepancies. Determining the cause of discrepancies is an important step as it may result in the identification of new locations of concern (e.g., storage areas or process lines have been moved to a different part of the plant). This process may also add/delete chemicals or other materials to/from the list being evaluated (e.g., where a chemical is no longer in use, or where a chemical substitution has been made).

The field evaluation also provides an opportunity to look for evidence of past releases or situations that represent potential releases to the environment. Notes should be assembled indicating the substances that might be released and the migration pathway that would be followed by any such release. This information should be correlated with the facility map. Where evidence of past leaks is found, further study should be undertaken to determine if the evidence correlates with the release information already obtained.

The third step in the release identification and assessment process involves evaluating potential pollutants based on the hazards they present to human health and the environment. No single measure of toxicity or hazardous exists because chemicals may have a variety of effects (both direct and indirect) that are characterized by a range of physical/chemical properties and associated effects. Some chemicals, for example, may be hazardous because of flammability and therefore represent fire hazards. Other products may be toxic and represent a threat to waterways and their associated flora and fauna, contaminate groundwaters, and/or threaten workers cleaning-up spills who are not provided with the proper protective equipment (e.g., respirators). Potential releases of pollutants to the environment might be subject to regulation under environmental permits, and represent threats to the facility in the form of noncompliance.

Detailed information on material properties should be available from plant safety personnel. When evaluating the threats posed by chemicals, facility personnel should consult available technical literature, manufacturer's representatives, and technical experts such as safety coordinators within the plant. A variety of technical resources can provide information of chemical properties including the following:

- Material safety data sheets
- American Council of Government and Industrial Hygienist publications on fume toxicity
- N. Sax, *Dangerous Properties of Industrial Material*, Seventh Edition, Volume 1-3, Van Nostrand Reinhold Company, Inc., New York, New York, 1989.
- *National Institute of Occupational Safety and Health (NIOSH) Pocket Guide to Chemical Hazards*, U.S. Department of Health and Human Services, 1990.
- EPA guidance documents. (Call EPA Public Information Center (202) 260-7751.)

These references can provide information on specific physical/chemical properties that should be considered in evaluating hazards, including toxicity, ignitability, explosivity, reactivity and corrosivity. Careful evaluation of these data will provide a basis for determining the intrinsic threat posed by materials at the facility. Armed with such understanding and subsequent identification of exposure pathways and potential receptors (the next step in the process), the need for developing BMPs comes into focus.

The fourth step in the release identification and assessment process involves identifying pathways by which pollutants identified at the site might reach environmental and human receptors. Identifying the pathways of current releases can easily be accomplished based on visual observations.

However, identifying the pathways of potential releases requires the use of sound engineering judgement in determining the point of release, estimating the direction and rate of flow of potential releases toward receptors of concern, and identification and technical evaluation of any existing means of controlling chemical releases or discharges (such as dikes or diversion ditches).

Information from the site map and observations made during the visual inspection (e.g., location of materials, potential release points, drainage patterns) should prove useful in this analysis. Of primary concern will frequently be exposures to workers in the immediate area of a release where concentrations will be highest. Migration pathways for other exposures will often be of secondary concern.

When identifying pathways and receptors, all logical alternative pathways should be considered. Contaminations may be released through a number of methods (e.g., volatilization, leaching, runoff) to a number of media (e.g., air, groundwater, soil), all which may result in release to water. The analyst should consider all pathways carefully in combination with the materials inventory to identify possible release mechanisms and receptor media.

During the site-assessment, each area should be evaluated for potential problems. These problems might include equipment failure, evidence of wear or corrosion, improper operation (e.g., a tank overflow or leakage or exposure of raw material to runoff), problems caused by natural conditions (e.g., cracks or joint separation due to extremes in temperature), and materials incompatibility. The adequacy of control and planned remedial measures should also be examined. For example, the volume of oil in a storage tank holding liquid petroleum fuel oil might exceed the amount that could be controlled by a dike or a berm in the case of a tank failure. Increasing the size of containment can remedy such a problem. The availability and location of absorbent materials and/or booms would be of interest in case of spill or tank failure and should be evaluated to determine sufficiency.

The fifth and final step in the release identification and assessment process requires the application of best professional judgment in prioritizing potential releases. Priorities should be established for both known and potential releases. A combination of information identified in the previous steps about releases (the probability of release, the toxicity or hazards associated with each pollutant, and descriptions of the potential pathways for releases) should be evaluated. Using this information, a facility can rank actual and potential sources as high, medium, or low priority. These priorities can then be used in developing a BMP plan that places the greatest emphasis on BMPs for the sources that present the greatest risk to human health and environment.

Release Potential Identification and Assessment - What to Do

- Consider using other resources when conducting the release identification and assessment. Corporate or brother/sister company personnel may be available for consultation and assistance. Also, non-regulatory onsite assistance may be available (see Chapter 4 of this manual for details).
- Utilize worksheets and boilerplate formats to ensure that information is organized, easily evaluated, and easily understood.
- Utilize videotapes and photos to capture a visual picture of the facility site for use in later assessment evaluations. These representations may also be useful in BMP plan evaluation/reevaluation.
- Consider conducting monitoring to identify pollutants, pollutant loadings, and sources.
- Conduct brainstorming sessions to gather creative solutions for prioritized problems, followed by screening to eliminate impractical resolution.
- Evaluate technical merits and economic benefits of alternatives in an organized fashion. Consider ranking alternatives based on effects to product quality, costs, environmental benefits, ease of implementation, and success in other applications.

Release Potential Identification and Assessment - What Not to Do

- Do not make the site map so busy that information cannot be discerned. Enlarge the site map, or separate information on transparencies to later superimpose on the base map.
- At large facilities, be cognizant of not overloading BMP committee members with release identification and assessment responsibilities. Consider the establishment of several evaluation teams, each assigned to assess a specific area.
- Do not make changes in processes prior to allowing for an update in the release identification and assessment. Allow for the determination as to whether alternate methodologies or materials can be identified which are more environmentally protective or cost effective.

- Do not proverbially bite off more than the company can chew. Consider implementing changes in stages. Simple, procedural changes can be implemented immediately, while evaluations may need to be performed prior to the adoption of other measures.

2.3.2 BMP Plan Development Phase

After the BMP policy statement and committee have been established and the release potential identification and assessment has defined those areas of the facility that will be targeted for BMPs, the committee can begin determining the most appropriate BMPs to control environmental releases. The BMP plan should consist of both facility-specific BMPs and general BMPs.

To provide a possible starting point in developing BMPs, Chapters 3 and 4 of this manual present industry-specific BMPs and resources available for determining BMPs, respectively. These chapters can be used as a convenient reference to determine one or more facility-specific BMPs that might serve to reduce, control, or eliminate site-, process-, or pollutant-specific releases of harmful substances. Facilities should select the most appropriate specific BMPs based on effectiveness in reducing, controlling, or eliminating pollutants and feasibility.

General BMPs are relatively simple to evaluate and adopt. As previously indicated, general BMPs are practiced to some extent at all facilities. It is EPA's belief that all BMP plans should consist of six basic components:

- ***Good housekeeping:*** A program by which the facility is kept in a clean and orderly fashion
- ***Preventive maintenance:*** A program focused on preventing releases caused by equipment problems, rather than repair of equipment after problems occur
- ***Inspections:*** A program established to oversee facility operations and identify actual or potential problems
- ***Security:*** A program designed to avoid releases due to accidental or intentional entry
- ***Employee training:*** A program developed to instill in employees an understanding of the BMP plan
- ***Recordkeeping and reporting:*** A program designed to maintain relevant information and foster communication.

A discussion of each of these basic components follows.

2.3.2.1 Good Housekeeping

What Is Good Housekeeping?

Good housekeeping is essentially the maintenance of a clean, orderly work environment. Maintaining an orderly facility means that materials and equipment are neat and well-kept to prevent releases to the environment. Maintaining a clean facility involves the expeditious remediation of releases to the environment. Together, these terms, clean and orderly, define a good housekeeping program.

Maintaining good housekeeping is the heart of a facility's overall pollution control effort. Good housekeeping cultivates a positive employee attitude and contributes to the appearance of sound management principles at a facility. Some of the benefits that may result from a good housekeeping program include ease in locating materials and equipment; improved employee morale; improved manufacturing and production efficiency; lessened raw, intermediate, and final product losses due to spills, waste or releases; fewer health and safety problems arising from poor materials and equipment management; environmental benefits resulting from reduced releases of pollution; and overall cost savings.

How Does a Good Housekeeping Program Function?

Good housekeeping measures can be easily and simply implemented. Some examples of commonly implemented good housekeeping measures include the orderly storage of bags, drums, and piles of chemicals; prompt cleanup of spilled liquids to prevent significant runoff to receiving waters; expeditious sweeping, vacuuming, or other cleanup of accumulations of dry chemicals to prevent them from reaching receiving waters; and proper disposal of toxic and hazardous wastes to prevent contact with and contamination of storm water runoff.

The primary impediment to a good housekeeping program is a lack of thorough organization. To overcome this obstacle, a three-step process can be used, as follows:

- Determine and designate an appropriate storage area for every material and every piece of equipment

- Establish procedures requiring that materials and equipment be placed in or returned to their designated areas

- Establish a schedule to check areas to detect releases and ensure that any releases are being mitigated.

The first two steps act to prevent releases that would be caused by poor housekeeping. The third step acts to detect releases that have occurred as a result of poor housekeeping. Exhibit 2-7 provides an example of a good housekeeping program that has functioned to prevent releases.

How Is a Good Housekeeping Program Created?

As with any new or modified program, the initial stages will be the largest hurdle; ultimately, though, good housekeeping should result in savings that far outweigh the efforts associated with initiation and implementation. Generally, a good housekeeping plan should be developed in a manner that creates employee enthusiasm and thus ensures its continuing implementation.

The first step in creating a good housekeeping plan is to evaluate the facility site organization. In most cases, a thorough release identification and assessment has already generated the needed inventory of materials and equipment and has determined their current storage, handling, and use locations. This information together with that from further assessments can then be used to determine if the existing location of materials and equipment are adequate in terms of space and arrangement.

Cramped spaces and those with poorly placed materials increase the potential for accidental releases due to constricted and awkward movement in these areas. A determination should be made as to whether materials can be stored in a more organized and safer manner (e.g., stacked, stored in bulk as opposed to individual containers, etc). The proximity of materials to their place of use should also be evaluated. Equipment and materials used in a particular area should be stored nearby for convenience, but should not hinder the movement of workers or equipment. This is especially important for waste products. Where waste conveyance is not automatic (e.g., through chutes or pipes) waste receptacles should be located as close as possible to the waste generation areas, thereby preventing inappropriate disposal leading to environmental releases.

EXHIBIT 2-7: AN EXAMPLE OF THE SUCCESSFUL IMPLEMENTATION OF A GOOD HOUSEKEEPING PROGRAM

Emerson Electric Company's Murphy, North Carolina location developed a good housekeeping program which resulted in better waste management. Emerson noted that a number of activities contributed unwanted pollutants to their treatment plant and storm water discharges. These included oil spills from the scrap loading site (20 gallons per week), spills from the aluminum die-casting operations (45 gallons per week), and dumping of miscellaneous chemicals including monthly dumping of alkaline cleaner. Additionally, the company noted that unlabelled hazardous chemicals were located in random locations, some in outside storage areas. These chemicals and other non-hazardous substances were identified as having the potential to result in spills of up to 20,000 pounds.

As a result, the facility established good housekeeping procedures and measures which included:

- Installation of sump and pump in the die-casting and scrap loading areas which recovered 65 gallons of oil per week.
- Requirements for the discontinuance of dumping activities.
- Implementation of labeling and manifesting procedures for all hazardous wastes and the storage of these wastes in inside controlled areas.

The program involved informing personnel of their responsibilities under the program and the maintenance of daily log sheets which demonstrate proper activities. The ongoing good housekeeping program is monitored closely by the in-plant process engineer. Any violations of good housekeeping procedures are reported to and addresses by the plant manager.

Adapted from: D. Huisingh, L. Hilger, N. Seldman, *Proven Profits from Pollution Prevention: Case Studies in Resource Conservation and Waste Reduction,* Institute for Local Self-Reliance, Washington, D.C., 1985.

Appropriately designated areas (e.g., equipment corridors, worker passageways, dry chemical storage areas) should be established throughout the facility. The effective use of labeling is an integral part of this step. Signs and adhesive labels are the primary methods used to assign areas. Many facilities have developed innovative labeling approaches, such as color coding the equipment and materials used in each particular process. Other facilities have stenciled outlines to assist in the proper positioning of equipment and materials.

Once a facility site has been organized in this manner, the next step is to ensure that employees maintain this organization. This can be accomplished through explaining organizational procedures to employees during training sessions (see 2.3.2.5 for information on training programs), distributing written instructions, and most importantly, demonstrating by example.

Support of the program must be demonstrated, particularly by responsible facility personnel. Shift supervisors and others in positions of authority should act quickly to initiate activities to rectify poor housekeeping. Generally, employees will note this dedication to the good housekeeping program and will typically begin to initiate good housekeeping activities without prompting. Although initial implementation of good housekeeping procedures may be challenging, these instructions will soon be followed by employees as standard operating procedures.

Despite good housekeeping measures, the potential for environmental releases remains. Thus, the final step in developing a good housekeeping program involves the prompt identification and mitigation of actual or potential releases. Where potential releases are noted, measures designed to prevent release can be implemented. Where actual releases are occurring, mitigation measures such as those described below may be required.

Mitigative practices are simple in theory: the immediate cleanup of an environmental release lessens chances of spreading contamination and lessens impacts due to contamination. When considering choices for mitigation methods, a facility must consider the physical state of the material released and the media to which the release occurs. Some considerations are provided in Exhibit 2-8. Generally, the ease of implementing mitigative actions should also be considered. For example, diet, crushed stone, asphalt, concrete, or other covering may top a particular area. Consideration as to which substance would be easier to clean in the event of a release should be evaluated.

Conducting periodic inspections is an excellent method to verify the implementation of good housekeeping measures. Inspections may be especially important in the areas identified in the release identification and assessment step where releases have previously occurred. Inspections and related concerns are discussed in Section 2.3.2.3.

EXHIBIT 2-8: CONSIDERATIONS FOR SELECTION OF MITIGATIVE PRACTICES

- Manual cleanup methods, such as sweeping and shoveling, are generally most appropriate for materials released in the solid phase to solid media and small releases of liquids which have saturated the soil.
- Mechanical cleanup methods such as excavation practices (e.g., plowing, backhoeing), are most appropriately used for large releases of solid phase materials to solid media and for larger areas contaminated by liquid material releases to the soil. Vacuum systems, a less common mechanical cleanup method, can be used for large releases of solid phase materials to solid media and for removing liquids released to water media when mixing has not occurred.
- Other cleanup methods may be the only option for mitigating the release of certain materials to the environment. These include the following:
  - Sorbents such as straw, sawdust, clay, activated carbon and miscellaneous complex organics may be used to clean up small gaseous or liquid releases to water and solid media. Sorbents must later be remediated by manual or mechanical cleanup methods.
  - Gelling agents including polyelectrolytes, polyacrylamide, butylstyrene copolymers, polyacrylonitrile, polyethylene oxide, and the universal gelling agent interacts with a liquid or gaseous releases to form a more viscous mass which can then be remediated by manual or mechanical cleanup methods. Gelling agents can effectively mitigate liquid releases prior to discharge to a water media or infiltration into the soil.

It may not always be possible to immediately correct poor housekeeping. However, deviations should occur only in emergencies. The routines and procedures established as a part of the program should allow for adequate time to conduct good housekeeping activities.

Good Housekeeping - What to Do

- Integrate a recycling/reuse and conservation program in conjunction with good housekeeping. Include recycle/reuse opportunites for common industry wastes such as paper, plastic, glass, aluminum, and motor oil, as well as facility-specific substances such as chemicals, used oil, dilapidated equipment, etc. into the good housekeeping program. Provide reminders of the need for conservation measures including turning off lights and equipment when not in use, moderating heating/cooling, and conserving water.

- When reorganizing, keep pathways and walkways clear with no protruding containers.

- Create environmental awareness by celebrating Earth Day (April 22) and/or developing a regular (e.g., monthly) good housekeeping day.

- Develop slogans and posters for publicity. Involve employees and their families by inviting suggestions for slogans and allowing children to develop the facility's good housekeeping posters.

- Provide suggestion boxes for good housekeeping measures.

- Develop a competitive program that may include company-wide competition or facility-wide competition. Implement an incentive program to spark employee interest (i.e., ½ day off for the shift which best follows the good housekeeping program).

- Conduct inspections to determine the implementation of good housekeeping. These may need to be conducted more frequently in areas of most concern.

- Pursue an ongoing information exchange throughout the facility, the company, and other companies to identify beneficial good housekeeping measures.

- Maintain necessary cleanup supplies (i.e., gloves, mops, brooms, etc.).

- Set job performance standards which include aspects of good housekeeping.

Good Housekeeping - What Not to Do

- Do not allow rubbish or other waste to accumulate. Properly dispose of waste, or arrange to have it removed in a timely fashion.

- Do not limit good housekeeping measures to industrial locations. Office areas should also be involved in the good housekeeping program.

2.3.2.2 Preventive Maintenance

What Is Preventive Maintenance?

Preventive maintenance (PM) is a method of periodically inspecting, maintaining, and testing plant equipment and systems to uncover conditions which could cause breakdowns or failures. As part of a BMP plan, PM focusses on preventing environmental releases. Most facilities have existing PM programs. It is not the intent of the BMP plan to require development of a redundant PM program. Instead, the objective is to have personnel evaluate their existing PM program and recommend changes, if needed, to address concerns raised as part of the release potential identification and assessment (See Section 2.3.1.3). Ultimately, this will result in the focus of preventive maintenance on the areas and pollutants determined to be of most concern. Where no refocussing is necessary, the PM program suggested as part of the BMP plan and the existing PM program can be identical.

A PM program accomplishes its goals by shifting the emphasis from a repair maintenance system to a preventive maintenance system. It should be noted that in some cases, existing PM programs are limited to machinery and other moving equipment. The PM program prescribed to meet the goals of the BMP plan includes all other items (man-made and natural) used to contain and prevent releases of toxic and hazardous materials. Ultimately, the well operated PM program devised to support the BMP plan should produce environmental benefits of decreased releases to the environment, as well as reducing total maintenance costs and increasing the efficiency and longevity of equipment, systems, and structures.

How Does a Preventive Maintenance Program Function?

In terms of BMP plans, the PM program should prevent breakdowns and failures of equipment, containers, systems, structures, or other devices used to handle the toxic or hazardous chemicals or wastes. To meet this goal, a PM program should include a suitable system for evaluating equipment, systems, and structures; recording results; and facilitating corrective actions. A PM program should, at a minimum, include the following activities:

- Identification of equipment, systems, and structures to which the PM program should apply
- Determination of appropriate PM activities and the schedule for such maintenance
- Performance of PM activities in accordance with the established schedule
- Maintenance of complete PM records on the applicable equipment and systems, and structures.

Generally, the PM program is designed to prevent and/or anticipate problems resulting from equipment and structural failures. However, it is unrealistic to expect that the PM program will avert the need for repair maintenance as a result of unanticipated problems. Adjustments and repair of equipment will still be necessary where problems occur, and replacement of equipment will be necessary when adjustment and repair are not sufficient.

Generally, all good PM programs will consist of the four components noted above. However, it is of particular importance that the PM program address those areas and pollutants identified during the release identification and assessment step. Exhibit 2-9 provides a summary of releases which are attributed in part to inadequate PM programs. These releases demonstrate the need for an effective PM program.

EXHIBIT 2-9: DEMONSTRATION OF THE NEED FOR AN EFFECTIVE PM PROGRAM

From a period beginning July 1983 to July 1988, EPA recorded the following catastrophic spills: 126,000 gallons crude oil; 8 tons anhydrous ammonia; 10,000 gallons hydrochloric acid; 100,000 gallons toluene, xylene and methyl-ethyl ketone; 1,000 bbl of phenol; 60,000 gallons of sodium hydroxide; 3,000 gallons or aromatic hydrocarbons; 50,000 pounds of phenol and cyclohexane; 3,000 gallons of miscellaneous solvent; 60,700 gallons of sodium bisulfite; 100,000 gallons of a combination of cadmium, phenol, and methylene chloride; 700,000 gallons of ammonium nitrate; 25,000 gallons of jet fuel; 8 tons of anhydrous ammonia; and 1,500 gallons of hexanol isobutyrate.

Adapted from: "Best Management Practices (BMPs) in NPDES Permits-Information Memorandum, "EPA Office of Water, dated April 15, 1983, March 23, 1984, June 3, 1985, August 29, 1986, August 11, 1987, and August 19, 1988.

How Is a Preventive Maintenance Program Created?

Although creating and implementing PM programs sounds easy, it is often impeded by lack of funding and of organization. Lack of funding must be overcome by a facility's commitment to its PM program based on the simple truth that PM is less costly than replacement. Lack of organization can be overcome by better planning, which can be achieved by following the steps to developing an effective PM plan discussed below.

At the outset of a PM program, an inventory should be devised. This inventory should provide a central record of all equipment and structures including: location; identifying information such as serial numbers and facility equipment numbers/names; size, type, and model; age; electrical and mechanical data; the condition of the equipment/structure; and the manufacturer's address, phone number, and person to contact. In addition to the equipment inventory, an inventory of the structures and other non-moving parts to which the PM program is to apply should also be determined.

Inventories can be developed through inspections and/or reviews of facility specifications and operations and maintenance manuals. In some cases, it is effective to label equipment and structure with assigned numbers/names and some of the identifying information. This information may be useful to maintenance personnel in the event of emergency situation or unscheduled maintenance where maintenance information is not readily available. Several different methods are effective for recording inventory information including the use of index cards, prepared forms and checklists, or a computer database.

Since the PM program involves the use of maintenance materials (i.e., spare parts, lubricants, etc.), some additional considerations may apply. First, good housekeeping measures, as discussed in Section 2.3.2.1, are particularly important for organizing maintenance materials and keeping areas clean. A tracking system may also be necessary for organizing maintenance materials. The inventory should include information such as materials/parts description, number, item specifications, ordering information, vendor addresses and phone numbers, storage locations, order quantities, order schedules and costs. A large facility may require a parts catalog to coordinate such information. Large facilities may also find it necessary to develop a purchase order system which maintains the stock in adequate number and in the proper order by keeping track of the minimum and maximum number of items required to make timely repairs, parts that are vulnerable to breakage, and parts that have a long delivery time or are difficult to obtain.

Once the inventory is completed, the facility should determine the PM requirements including schedules and specifications for lubrication, parts replacement, equipment and structural testing, maintenance of spare parts, and general observations. The selected PM activities should be based on the facility-specific conditions but should be at least as stringent as the manufacturer's recommendations. Manufacturer's specifications can generally be found in brochures and pamphlets accompanying equipment. An operations and maintenance manual also may contain this information. If these sources are not available, the suggested manufacturer's recommendation can be obtained directly from the manufacturer. In cases of structures or non-moving parts, the facility will need to determine an appropriate maintenance activities (e.g., integrity testing). As with inventory information, PM information should be recorded in an easily accessible format.

After establishment of the materials inventory and the development of PM requirements, a facility should schedule and carry out PM on a regular basis. Personnel with expertise in maintenance should be available to conduct maintenance activities. In a small facility where one person may conduct regular maintenance activities, specialized contractors may supplement the maintenance program for more complex activities. An up-to-date list of outside firms available for contract work beyond the capability of the facility staff should be readily available. Additionally, procedures explaining how to obtain such support should be provided in the BMP plan. Larger facilities should have sufficient PM expertise within the staff including a PM manager, an electrical supervisor, a mechanical supervisor, electricians, technicians, specialists, and clerks to order and acquire parts and maintain records. Ongoing training and continuing education programs may be used to establish expertise in deficient areas. Training is discussed in more detail in Section 2.3.2.5.

Maintenance activities should be coordinated with normal plant operations so that any shut-downs do not interfere with production schedules or environmental protection. Examples showing the results of poor coordination between operations and maintenance staff are given in Exhibit 2-10.

EXHIBIT 2-10: EXAMPLES OF POOR COORDINATION BETWEEN OPERATIONS AND MAINTENANCE STAFF

Operator-caused	Inattention to unusual noises or conditions; too many motor stops and starts in one day; tampering with limit switches; failure to report suspected problems.
Maintenance-caused	Not replacing trip switches; bypassing fail safe systems or instrumentation; imprecise equipment alignment; failure to report other noticed equipment deficiencies

Adapted from: *Plant Maintenance Program Manual of Practice OM-3*, Water Pollution Control Federation, Alexandria, VA, 1982.

The maintenance supervisory staff should also consider other timing constraints such as the availability of the PM staff for both regularly scheduled PM and unanticipated corrective repairs.

The final step in the development of a PM program involves the organization and maintenance of complete records. A PM tracking system which includes detailed upkeep, cost, and staffing information should be utilized. A PM tracking system assists facilities in: identifying potential equipment or structural problems resulting from defects, general old age, inappropriate maintenance, or poor engineering design; preparation of a maintenance department budget; and deciding whether a piece of equipment or a structure should continue to be repaired or replaced.

There are many commercial software systems that enable facilities to track maintenance. Computer systems allow for input of inventory and PM information and generate daily, weekly, monthly, and/or yearly maintenance sheets which include the required the item to be maintained, the maintenance duties, and materials to be used (e.g., oil, spare parts, etc.). The system can be continually updated to add information gathered during maintenance activities. Some of the maintenance information that proves useful includes the work hours spent, materials used, frequency of downtime for repairs, and costs involved with maintenance activities. This information in turn can generate budgets and determinations of the cost effectiveness of repair versus replacement, etc. Computerized systems for maintenance tracking are usually most effective at larger facilities.

Useful manual systems may involve index cards, maintenance logs, and a maintenance schedule. Initially, inventory and PM information can be recorded on index cards. This information can be consulted during maintenance activities. Maintenance logs should also be developed for each piece of equipment and each structure, and should contain information such as the maintenance specifications, and data associated with the completion of maintenance activities. Maintenance personnel should complete relevant information including the date maintenance was conducted, hours spent on duties, materials used, worker identification, and the nature of the problem. Appendix B provides some examples of formats to use in organizing and recording inventory, PM requirements, and PM duty information.

Preventive Maintenance - What to Do

- When attaching information to equipment and structures, use bold and bright colors consistent with the approaches described for good housekeeping.
- Consider and discuss additional PM procedures beyond those normally recommended by the manufacturer.
- Conduct extensive safety training for PM personnel.
- Coordinate scheduling of PM activities with facility or unit downtime.
- Keep track of how long materials have been stored. This will support an evaluation of the integrity of storage containers.
- Develop a PM staff team approach including team names (i.e., the A-team) to create enthusiasm.
- Utilize blackboards and charts to assist in organizing and conveying an annual PM schedule.

Preventive Maintenance - What Not to Do

- Do not forget to stock important replacement parts and any specialized tools required to repair equipment.
- Do not create a paperwork nightmare. Develop the minimum number of well-organized logs necessary to maintain information.
- Do not let untrained, unskilled personnel conduct PM activities. Employees taking part in the PM program must be familiar with equipment and maintenance procedures.
- Do not forget to determine the availability and time needed to obtain vital parts or contractor assistance.

2.3.2.3 Inspections

What Are Inspections?

Inspections provide an ongoing method to detect and identify sources of actual or potential environmental releases. For example, Exhibit 2-6 in Section 2.3.1.3 described the use of an inspection during release identification and assessment. Inspections also act as oversight mechanisms to ensure that selected BMPs are being implemented. Inspections are particularly effective in evaluating the good housekeeping and PM programs previously discussed. Many of the releases highlighted in Exhibit 2-9 in Section 2.3.22 also can be partly attributed to failures in inspection programs. In addition, Exhibit 2-11 describes a number of additional releases which resulted primarily from ineffective inspection programs.

EXHIBIT 2-11: RELEASES WHICH COULD HAVE BEEN PREVENTED BY EFFECTIVE INSPECTION PROGRAMS

From July 1983 to July 1988, EPA recorded the following releases: 100,000 gallons of toluene, xylene, and methyl ethyl ketone; 300 gallons of phosphorous trichloride; 1,100 gallons of trichloroethane; 3,000 gallons of heavy polymer distillate; more than 8,000 gallons of methyl isobutyl ketone; and 100,000 gallons of a combination of cadmium, phenol, and methylene chloride.

Adapted from: *Best management Practices (BMPs) in NPDES Permits - Information Memorandum*, EPA Office of Water, dated April 15, 1983, March 23, 1984, June 3, 1985, August 29, 1986, August 11, 1987, and August 19, 1988.

Many facilities may be currently conducting inspections, but in a less formalized manner. Security scans, site reviews, and facility walk throughs conducted by plant managers and other such personnel qualify as inspections. These types of reviews, however, are often limited in scope and detail. To ensure the objectives of the BMP plan are met, these types of reviews should be conducted concurrently with periodic, in-depth inspections as part of a comprehensive inspection program.

How Does an Inspection Program Function?

Inspections implemented as part of the BMP plan should cover those equipment and facility areas identified during the release identification and assessment as having the highest potential for environmental releases. Since inspections may vary in scope and detail, an inspection program should be developed to prevent redundancy while still ensuring adequate oversight and evaluation.

A BMP inspection program should set out guidelines for: (1) the scope of each inspection; (2) the personnel assigned to conduct each inspection; (3) the inspection frequency; (4) the format for reporting inspection findings; and (5) remedial actions to be taken as a result of inspection findings.

Despite the different requirements of each type of inspection, the focus of inspections conducted as part of the BMP plan should not vary. As discussed in Section 2.2, some of the areas within the facility that may be the focus of the BMP plan include solid and liquid materials storage areas, in-plant transfer and materials handling areas, activities with potential to contaminate storm water runoff, and sludge and hazardous waste disposal sites.

How Is an Inspection Program Created?

An inspection program's goal will be to ensure thoroughness, while preventing redundancy. Ultimately, this will ensure that the use of resources is optimized. In addition, it should be clear that the inspection team's efforts are directed to support the operating groups in carrying out their responsibilities for equipment and personnel safety, and work quality, and to ensure that all standards are met. In achieving these goals, written procedures discussing the scope, frequency and scheduling, personnel, format, and remediation procedures should be provided. These are discussed in the following paragraphs.

The scope of each inspection type should be discussed in the written procedures. Many different types of inspections are conducted as part of the inspection program. Guidelines for the scope of these inspections include:

- ***Security scan***: Search for leaks and spills which may be occurring. Specifically examine problems areas which have been identified by the plant manager or equivalent persons.

- ***Walk through***: Conduct oversight of the duties associated with a security scan. In addition, ensure that equipment and materials are located in their appropriate positions.

- ***Site review***: Conduct oversight of duties associated with a walk through. Additionally, evaluate the effectiveness of the PM, good housekeeping, and security programs by visual oversight of their implementation.

- ***BMP plan oversight inspection***: Conduct oversight of duties associated with a site review. Evaluate the implementation of all aspects of the written BMP plan including the review of the records generated as part of these programs (e.g., inspection reports, PM activity logs).

- ***BMP plan evaluation/reevaluation inspection***: Conduct an evaluation/reevaluation of the facility and determine the most appropriate BMPs to control environmental releases.

An appropriate mix of these types of inspections should be developed based on facility-specific considerations. The proper frequency for conducting inspections will vary based on the type of the inspection and other facility-specific factors. Some general guidelines for establishing frequency follow:

- Security scans can be conducted various times daily
- Walk through inspections can be conducted once per shift to once per week
- Site reviews can be conducted once per week to once per six months
- BMP plan oversight inspections can be conducted once per month to once per year
- BMP plan reevaluation inspections can be conducted once per year to once every five years.

There are no hard and fast rules for conducting inspections as part of the BMP plan. Inspection frequencies should be based on a facility's needs. Two points should be considered when establishing an inspection program: (1) As would be expected, more frequent inspections should be conducted in the areas of highest concern; and (2) inspections must be conducted more frequently during the initial BMP implementation until the BMP plan procedures become part of standard operating procedures.

It may be useful to set up a schedule to ensure a comprehensive inspection program. Varying the dates and times of inspection conduct is also good practice in that it ensures all stages of production and all situations are reviewed.

Individuals qualified to assess the potential for environmental releases should be assigned to conduct formal inspections. Members of the BMP committee can generally fulfill this requirement, but they may not be available to conduct all inspections. Thus, it may be appropriate to identify and train personnel to conduct specific types of inspections. For example, shift foremen and other equivalent supervisory personnel may appropriately conduct walk throughs and site reviews as a result of their position of authority and ability to require prompt correction if problems are observed. Personnel with immediate responsibility for an area should not be asked to conduct inspections of that area as they may be tempted to overlook problems. Additionally, plant security and other personnel who routinely conduct walk throughs should not be assigned to conduct BMP plan inspections since their familiarity with the facility may result in their not being suited to best identify opportunities for improvement.

Different perspectives are useful when conducting inspections. By developing a team inspection approach or by alternating inspectors, facilities can receive a more thorough review. One inspector may observe something that another will overlook, and an inspector tends to focus on the areas with which he/she is most familiar.

An inspection checklist of areas to inspect with space for a narrative report is a helpful tool when conducting inspections. A standard form helps ensure inspection consistency and comprehensiveness. Sample inspection forms and checklists are provided in Appendix B.

Checklists may not be necessary for each inspection performed. This may be particularly true for facilities conducting frequent inspections (once per hour, once per shift, etc.); procedures for using inspection checklists should be reasonable to prevent paperwork nightmares.

The findings of inspections will be useless unless they are brought to the attention of appropriate personnel and subsequently acted upon. To ensure that reports are acted upon in an expeditious and appropriate manner, procedures for routing and review of reports should be developed and followed. Recordkeeping and reporting is discussed in Section 2.3.2.6.

Despite the usefulness of written reports, in no way should a written report replace verbal communication. Where a problem is noted, particularly environmental releases currently occurring or about to occur, it should be verbally communicated by the inspector to the responsible personnel as soon as possible.

Inspections - What to Do

- Encourage workers to conduct visual inspections and report any actual or potential problems to the appropriate personnel.
- Develop inspection checklists for each type of inspection. Vary them where necessary for each part of the facility subject to BMPs.
- Consider utilizing non-regulatory support from EPA, States, or university supported resources when conducting site assessments.

Inspections - What Not to Do

- Do not rely solely on the use of a checklist for inspections. Narrative descriptions should be included in the reports to ensure that problems are identified and discussed.
- Do not conduct inspections and then fail to provide feedback of findings of concern to the person responsible for the area inspected.

2.3.2.4 Security

What Is a Security Plan?

A security plan describes the system installed to prevent accidental or intentional entry to a facility that might result in vandalism, theft, sabotage, or other improper or illegal use of the facility. In relation to a BMP plan, a security system should prevent environmental releases caused by any of these improper or illegal acts.

Most facilities already have a program for security in place; this security program can be integrated into the BMP plan with minor modifications. Facilities developing a program for security as part of the BMP plan may be hesitant to describe their security measures in detail due to concerns of compromising the facility. The intent of including a security program as part of the BMP plan is not to divulge facility or company secrets; the specific security practices for the facility may be kept as part of a separate confidential system. The security program as part of the BMP plan should cover security in a general fashion, and discuss in detail only the practices which focus on preventing environmental releases.

How Does a Security Plan Function?

The security program as part of the BMP plan should be designed to meet two goals. First, the security plan should prevent security breaches that result in the release of hazardous or toxic chemicals to the environment. The second goal is to effectively utilize the observation capabilities of the security plan to identify actual or potential releases to the environment. Some of the components that are typically included in a security scan are provided in Exhibit 2-12.

EXHIBIT 2-12: POSSIBLE COMPONENTS OF A SECURITY PLAN

- Routine patrol of the facility property by security guards in vehicles or on foot
- Fencing to prevent intruders from entering the facility site
- Good lighting to facilitate visual inspections at night, and of confined spaces
- Vehicular traffic control (i.e., signs)
- Access control using guardhouse or main entrance gate, where all visitors and vehicles are required to sign in and obtain a visitor's pass
- Secure or locked entrances to the facility
- Locks on certain valves or pump starters
- Camera surveillance of appropriate sites, such as facility entrance, and loading and unloading areas
- Electronic sensing devices supplemented with audible or covert alarms
- Telephone or other forms of communication.

How Is a Security System Created?

Typically, security systems focus on the areas with the greatest potential for damage as a result of security breaches. As part of the BMP plan, the security program will focus on the areas that result in environmental releases. Typically, these areas have been identified in the release identification and assessment step. In many cases, the findings of this step may indicate a need to change the focus or broaden the scope of the security program to include areas of the facility addressed by the BMP plan. Since the security program may not be common knowledge, general BMP committee members may not be able to recommend changes. As a result, security personnel should be involved in the decisions made by the committee, with one person possibly serving as a member.

While performing their duties, security personnel can actively participate in the BMP plan by checking the facility site for indications of releases to the environment. This may be accomplished by checking that equipment is operating properly; ensuring no leaks or spills are occurring at materials storage areas; and checking on problem areas (i.e., leaky valves, etc).

The advantages of integrating security measures into the BMP plan are considerable. Security personnel are in positions that enable them to conduct periodic walk throughs and scans of the facility, as well as covertly view facility operations. They are in an excellent position to identify and prevent actual or potential releases to the environment.

Where security personnel are utilized as part of the oversight program, two obstacles generally must be overcome: (1) support must be gained from the security staff; and (2) security personnel must be knowledgeable about what may and may not be a problem, and to whom to report when there is a problem. Involving the security staff in the BMP plan development at an early stage should assist in gaining their support. Integration of the security staff into the training, and recordkeeping and reporting programs discussed in Section 2.3.2.5 and 2.3.2.6, respectively, can also be used to overcome these barriers.

Security - What to Do

- File detailed documentation of the security system separately from the BMP plan to prevent unauthorized individuals from gaining access to confidential information.
- Make certain that all security personnel are aware of their assigned responsibilities under the BMP plan.
- Post security and informational signs and distribute security and direction information to visitors. This may be particularly useful for frequently visited buildings.

Security - What Not to Do

- Do not assume that isolation is adequate security.
- Do not locate alarms or indicator lights where they cannot be readily seen or heard.

2.3.2.5 Employee Training

What Is Employee Training?

Employee training conducted as part of the BMP plan is a method used to instill in personnel, at all levels of responsibility, a complete understanding of the BMP plan, including the reasons for developing the plan, the positive impacts of the plan, and employee and managerial responsibilities under the BMP plan. The employee training program should also educate employees about the general importance of preventing the release of pollutants to water, air, and land.

Training programs are a routine part of facility life. Most facilities conduct regular employee training in areas including fire drills, safety, and miscellaneous technical subject areas. Thus, the training program developed as a result of the BMP plan should be easily integrated into the existing training program.

Employee training conducted as part of the BMP plan should focus on those employees with direct impact on plan implementation. This may include personnel involved with manufacturing, production, waste treatment and disposal, shipping/receiving, or materials storage; areas where processes and materials have been identified as being of concern; and PM, security, and inspection programs. Training programs, which include all appropriate personnel, should include instruction on spill response, containment, and cleanup. Generally, the employee training program should serve to improve and update technical, managerial, or administrative skills; increase motivation; and introduce incentives for BMP plan implementation.

How do Employee BMP Training Programs Function?

Employee training programs function through: (1) analyzing training needs; (2) developing appropriate training materials; (3) conducting training; and (4) repeating training at appropriate intervals in accordance with steps 1 through 3. This four step process should be utilized during all employee training. An example of an employee training program is provided in Exhibit 2-13.

EXHIBIT 2-13: EXAMPLE OF THE IMPLEMENTATION OF EMPLOYEE TRAINING TO ENSURE THE SUCCESS OF ENVIRONMENTAL CONTROLS

The largest metal finishing shop in New England began a series of experiments with water reuse and conservation measures to reduce water usage from their high level of 140,000 gallons per day. Their studies resulted in the installation of flow nozzles which initially reduced their water use to 60,000 gallons per day. After reaching this level, experiments were discontinued. Immediately, water use rebounded to 100,000 gallons per day due to employee backsliding to previous less conservation-oriented practices.

Realizing that additional measures were necessary, the company installed dead rinse tanks and a series of countercurrent rinse tanks. These measures, supplemented by a program of recycle and reuse, reduced pollutant discharge concentration such that treatment was unnecessary to comply with effluent limitation. Additionally, the flow rate was reduced 40,000 gallons per day. However, due to their earlier experience with employee backsliding after implementation, the facility developed an employee training program that ensured a proper understanding of equipment operation and emphasized the benefits of water conservation. This maintained the impetus of the source reduction and recycling measures.

Adapted from: *Cutting the Cost of Complying with Electroplating Water Regulations Through Conservation*, EPA, 1982.

How is an Employee Training Program Created?

The first stage in developing a training program is analyzing training needs. Generally, training needs to be conducted during the planning and development phases of the BMP plan, and as follow-up to BMP implementation for selected areas of concern. In all three cases, it is important to analyze training needs and develop appropriate training tools to use during conduct of the training.

The initial BMP development session educates employees of the need for, objectives of, and projected impact of the BMP plan. As would be expected, this initial training should be conducted at the onset of the BMP development. The message portrayed at this session should be the positive impacts of the BMP plan including ease in locating materials and equipment; improved employee morale; improved manufacturing and production efficiency; lessened raw, intermediate and final product losses due to releases; fewer health and safety problems arising from unmitigated releases and/or poor placement of materials and equipment; environmental benefits resulting from reduced releases of pollution; and overall cost savings. When providing this message, it is essential that the benefits for employees as well as the company itself be stressed.

While it is important to point out the reasons that lead to the decision to implement a BMP plan, it is also important to provide a realistic picture of the changes and impacts which will result. These modifications should be discussed in terms of their positive impact to help maintain a high level of enthusiasm.

After the BMP plan is developed, the BMP implementation training sessions should be developed. The training sessions should review the BMP plan and associated procedures, such as the following:

- Any of the industry-specific BMPs selected from examples in Chapter 3 or developed based on facility-specific considerations
- The good housekeeping program including the use of labeling (signs, color coding, stenciling, etc.) to assign areas and procedures to return materials to assigned areas
- The PM program, including new PM schedules and procedures
- Integration of the security plan with the BMP plan
- Inspection programs
- Responsibilities under the recordkeeping and reporting system.

In some cases, it may be appropriate to provide a general session explaining BMP plan implementation followed by specialized training for each area. For example, since all employees should be aware of the good housekeeping program, this program should be discussed at the general session. Training for the selected facility-specific BMPs may be necessary only for employees in the production and manufacturing areas. PM information could be presented only to the personnel conducting maintenance, while security personnel need only be briefed of security-related responsibilities under the BMP plan.

Targeting the audience and determining training needs dictate many of the remaining aspects of employee training, including the following elements:

- Determination of meeting room sizes which will seat the audience comfortably.
- Selection of speakers qualified to discuss the topics to be presented.
- Determination of seating arrangement to best accomplish the goal of the training. Classroom style seating is best for lecture type sessions, while round tables will stimulate interactive type sessions.
- Selection of audiovisual aids such as podium mikes, lavaliere mikes, blackboards, standing overhead and/or slide projectors, video cassette recorders, and television monitors.
- Development of materials that convey training session information in a highly readable, yet creative format.
- Development of agendas that require consideration of all topics to ensure that the chosen topics can be discussed in the time allotted.
- Determination of training materials which must be prepared.

Training sessions are only as effective as the level of preparation. It is vital that workshop materials are technically accurate, easily read, and well organized. More importantly, training materials must leave a strong impression such that their message is remembered and any distributed training materials are consulted in the future. The use of audiovisual aids supplemented with informational handouts is one of the best methods of conveying information. Including copies of any slide or overhead helps avoid distractions during presentation caused by employees' copying contents of overheads. Other techniques which assist in effectively conveying information include the following:

- Providing aesthetically pleasing covers and professional looking handouts
- Developing detailed tables of contents with well numbered pages
- Frequently assimilating graphics into presentations
- Integrating break-out sections and exercises
- Incorporating team play during exercises
- Allowing for liberal question/answer sessions and discussions during or after presentations

- Providing frequent breaks
- Integrating field activities with class room training.

The use of qualified personnel to conduct training presentations also supports the facility's commitment to BMP plan implementation. Speakers should be identified in the initial training preparation stages based on their expertise in the topics to be presented. However, expertise is not the only consideration. Expertise must be supplemented with a well executed, interesting, enthusiastic presentation. Preparation prior to the training event will allow speakers to organize presentations, establish timing, and develop tone and content. Speakers should consider undergoing a dry run during which the speaker provides the full presentation including use of audio/visual aids.

Proper planning should ensure the execution of an effective training event. Once the training event has been conducted, some follow-up activities should be conducted. For example, evaluation forms requesting feedback on the training should be distributed to employees. These evaluation forms can be used to identify presentation areas needing improvement, ideas needing clarification, and future training activities. Ultimately, information gathered from these forms can help direct the employee training program in the future.

Once BMP plan implementation is underway, training should be conducted both routinely and on an as need basis. Special training sessions may also be prompted when new employees are hired, environmental release incidents occur, recurring problems are noted during inspections, or changes in the BMP plan are necessary.

Employee Training - What to Do

- Show strong commitment and periodic input from top management to the employee training program to create the necessary interest for a successful program.
- Ensure that announcements of training events are posted well in advance and include the times and dates of the sessions, the names and positions of the instructors, the lesson plans, and the subject material covered.
- Make the training sessions interesting. Use film and slide presentations. Bring in speakers to demonstrate the use of cleanup materials or equipment. Contact the State

health and environmental agencies or the regional EPA office for films, volunteer speakers, or other training aids.

- Use employee incentive programs or environmental excellence awards to reinforce training programs.
- Conduct demonstrative hands-on field training to show the effectiveness of good housekeeping, PM, or inspection programs.
- Give frequent refresher courses and consider pop quizzes to keep employees sharp.

Employee Training - What Not to Do

- Do not provide training to permanent facility employees only. Overlooking temporary and contractor personnel can increase the possibility of environmental releases.
- Do not allow training session attendance to be optional. Employees in the positions that incur the most stress in terms of meeting schedules should be reminded to avoid taking shortcuts when handling toxic or hazardous chemicals.
- Do not become too standardized. Reusing an annual employee training session will be tedious to employees. Integrate new information and improve on old information.

2.3.2.6 Recordkeeping and Reporting

What Is Recordkeeping and Reporting?

As part of a BMP plan, recordkeeping focusses on maintaining records that are pertinent to actual or potential environmental releases. These records may include the background information gathered as part of the BMP plan, the BMP plan itself, inspection reports, PM records, employee training materials, and other pertinent information.

Maintenance of records is ineffective unless a program for the review of records is set forth. In particular, a system of reporting actual or potential problems to appropriate personnel must be included. Reporting, as it relates to the BMP plan, is a method by which appropriate personnel are kept informed of BMP plan implementation such that appropriate actions may be determined and expeditiously taken. Reporting may be verbal or follow a more formal notification procedure. Some examples of reporting include the following:

- Informational memos distributed to upper management or employees to keep them updated on the BMP plan
- Verbal notification by BMP inspectors to supervisors concerning areas of concern noted during inspections
- Corrective action reports from the BMP committee to the plant manager which cite deficiencies with BMP plan implementation
- Verbal and written notification to regulatory agencies of releases to the environment.

How Does a Recordkeeping and Reporting Program Function?

A recordkeeping and reporting program effectively functions through the following three step procedure:

- Developing records in a useful format
- Routing records to appropriate personnel for review and determination of actions to address deficiencies
- Maintaining records for use in future decision making processes.

Recordkeeping and reporting play an overlapping role with the programs previously discussed. In general, these programs will involve the development, review, maintenance, and reporting of information to some degree. For example, an inspection program may include the development and use of an inspection checklist, submittal of the completed checklist to relevant personnel, evaluation of the inspection information, and determination of appropriate corrective actions. This may, in some cases, involve the development of a corrective action report to submit to appropriate persons (which may include regulatory agencies where necessary/required). The checklist and the corrective action reports should be maintained in organized files.

As part of the BMP plan, a recordkeeping and reporting program will primarily be developed for the PM (Section 2.3.2.2) and inspection (Section 2.3.2.3) programs. However, effective communication methods can also be useful in the development of the release identification and assessment portion of the BMP plan as discussed in Exhibit 2-14. A discussion of the step-by-step procedure for the development of an effective recordkeeping and reporting program follows.

EXHIBIT 2-14: EXAMPLE OF AN EFFECTIVE REPORTING PROGRAM DESIGNED TO PREVENT ENVIRONMENTAL RELEASES

West Point Pepperell, a textile manufacturing organization, established a Toxic Chemicals Committee consisting of a medical doctor, an industrial hygienist, three research chemists, an engineer, an attorney, a safety officer, a production representative, and an information specialist. The committee is designed to review chemical use at 40 manufacturing facilities across the country. Their reviews evaluate the chemical use danger to human health and the environment, and the availability of alternative chemicals.

The committee set forth notification procedures aimed at controlling new chemicals prior to the advent of use. These procedures required facilites to report potential chemical use to the committee for prior review and approval. Through these internal reporting procedures, the committee has been kept abreast of potentially discharged pollutants. Recommendations from the committee have resulted in the rejection of requests to use materials with potential to form benzidenes and bis(chloromethyl)ether. Additionally, the company attributes low levels of hazardous waste production to the work of the committee.

Adapted from: D. Huisingh, L. Martin, N. Hilger, N. Seldman, *Proven Profits from Pollution Prevention: Case Studies in Resource Conservation and Waste Reduction*, Institute for Local Self-Reliance, Washington, D.C., 1985.

How Is a Recordkeeping and Reporting Program Created?

A recordkeeping and reporting program must be developed in an organized manner. This ensures both the efficient use of resources and the compliance with regulatory requirements. Thus, a three step procedure involving development, review and reporting, and maintenance of records is suggested.

The first step to ensuring an effective recordkeeping and reporting program is the development of records in a useful format. The use of standard formats (i.e., checklists) can help to ensure the completion of necessary information, thoroughness in reviews, and understanding of the supplied data. For example, a standard inspection format may specify a summary of findings, recommendations, and requirements on the first page; then, detailed information by geographical area (e.g., materials storage area A, materials storage area B, the north loading and unloading zone) may be discussed. With a standard format, an inspection report reviewer may quickly review the findings summary to determine where problems exist, then refer to the detailed discussion of areas of concern. Ultimately, the use of a standard format minimizes the review time, expedites decisionmaking concerning corrective actions, and simplifies reporting.

Despite the recommended use of standard formats, inspectors should not feel constrained by the format. Sufficient detail must be provided in order for the report to be useful. Narratives should accompany checklists where necessary to provide detailed information on materials that have been released or have the potential to be released; nature of the materials involved; duration of the release or potential release; potential or actual volume; cause; environmental results of potential or actual releases; recommended countermeasures; people and agencies notified; and possible modifications to the BMP plan, operating procedures, and/or equipment.

The second step to ensuring an effective recordkeeping and reporting system involves routing information to appropriate personnel for review and determination of actions to address deficiencies. Regardless of whether the system for recordkeeping and reporting is structured or informal, the BMP plan should clearly indicate: (1) How information is to be transferred (i.e., by checklist, report, or simply by verbal notification); and (2) to whom the information is to be transferred (i.e., the plant manager, the supervisor in charge, or the BMP committee leader).

Customarily, formal means to transfer information would be more appropriate in larger more structured companies. For example, reviews of findings and conclusions as part of inspection reports may be conducted by supervisory personnel and the information may be routed through the chain of command to the responsible personnel such as shift supervisors or foremen. Less formal communication methods such as verbal notification may be appropriate for smaller facilities.

The key to ensuring a useful communication system is identifying one person (or, at larger facilities, several persons) to receive and dispense records and information. This person will be responsible for ensuring that designated individuals review records where appropriate, that corrective actions are identified, and that appropriate personnel are notified of the need to make corrections.

Additionally, this person will ensure that information is maintained on file for use in later evaluations of the BMP plan effectiveness.

It should be noted that the recordkeeping and reporting system is designed to help, not hinder, the communications process. Verbal communications of impending or actual releases should be made regardless of whether a formal communications process has been set forth.

A communications system for notification of potential or actual release should be designated. Such a system could include telephone or radio contact between transfer operations, and alarm systems that would signal the location of a chemical release. Provisions to maintain communication in the event of a power failure should be addressed. Reliable communications are essential to expedite immediate action and countermeasures to prevent incidents or to contain and mitigate chemicals released.

A reporting system should include procedures for notifying regulatory agencies. A number of agencies may require reporting of environmental releases including, Department of Transportation, Department of Energy, miscellaneous Department of Interior agencies, and the EPA. It is outside the scope of this manual to provide a summary of all necessary reporting requirements. However, reporting requirements specified under the NPDES permitting program include, at a minimum, the following:

- Releases in excess of reportable quantities which are not authorized by an NPDES permit
- Planned changes which
  - subject the facility to new source requirements
  - significantly change the nature or quantity of pollutants discharged
  - change a facility's sludge use or disposal practices
  - may result in noncompliance
- Notification within 24 hours of any unanticipated discharges (including bypasses and upsets) which may endanger human health or the environment, and the submission of a written report within five days
- The discharge of any toxic or hazardous pollutant above notification levels.
- Any other special notification procedure or reporting requirement specified in the NPDES permit.

Reports maintained in the recordkeeping system can be used in evaluating the effectiveness of the BMP plans, as well as when revising the BMP plan. Additionally, these records provide an oversight mechanism which allows the BMP committee to ensure that any detected problem has been adequately resolved. As such, the final step in developing a recordkeeping and reporting program involves the development and maintenance of an organized recordkeeping system.

In general, an organized filing system involves selecting an area for maintaining files, labelling files appropriately, and filing information in an organized manner. A single location should be designated for receiving the data generated for and related to the BMP plan. At larger facilities, several locations may be appropriate (e.g., maintenance records in one location, other BMP related documentation in another). A centralized location will help to consolidate materials for later review and consideration. Without a designated location, materials may become dispersed throughout a facility and subsequently lost.

Filing information by subject and date is a practice followed by most facilities. The most effective filing system usually includes hard copies of the information on a file. Additionally, keeping inventory lists of documents maintained in file folders assists in quick reviews of file contents. Small facilities may be able to file all BMP-related information in the same folder in chronological order; larger facilities may have to file information by subject. For example, PM information may be filed by equipment type in separate folders, while good housekeeping information and related oversight and evaluation information may be filed based on facility area. In some cases, larger facilities may find it convenient to develop an automated tracking system (e.g., a database system) for efficiently maintaining records.

Recordkeeping and Reporting - What to Do

- Clearly designate review and filing responsibilities for BMP related materials.
- Designate a file copy of any BMP correspondence.
- Set up procedures for materials release notification that include those plant personnel to be immediately notified, in order of priority, including backups, and then the appropriate governmental regulating agencies (Federal, State, and local). Include the fire department,

police, public water supply agency, fish and wildlife commission, and municipal sewage treatment plant, where appropriate.

- Develop a standard form for submitting a report for and the internal review of a release or near release.

- Share knowledge gained through BMP implementation with others. Report successes of BMP plan implementation in the Pollution Prevention Information Clearinghouse (see Chapter 4), magazines, or corporate newsletters.

Recordkeeping and Reporting - What Not to Do

- Do not keep the details of a materials release a secret known only to the facility management. Share the information learned from incidents so that all employees may benefit from the experience.

- Do not forget to keep employees informed. Continually provide updates (e.g., quarterly memo, newsletter) of BMP committee initiatives and progress. Lack of communication with employees may be interpreted as lack of continuing interest in the BMP plan's implementation.

2.3.3 BMP Plan Evaluation and Reevaluation

2.3.3.1 Plan Evaluation

Planning, development, and implementation of the BMP plan require the dedication of important resources by company management. The benefits derived, however, serve to justify the costs and commitments made to the BMP plan. To illustrate the plan's benefits, it may be appropriate and even necessary in some cases to measure the plan's effectiveness.

An evaluation can be performed by considering a number of variables, including: (1) benefits to the employees; (2) benefits to the environment; and (3) reduced expenditures. Methods of measuring these areas are discussed below.

Benefits to the employees can be assessed in terms of health and safety, productivity, and other factors such as morale. Comparisons before and after plan implementation can be made to determine trends that show BMP plan effectiveness. The following information can be utilized in this determination:

- Time off due to on-the-job injury or illness resulting from exposure to chemicals
- Production records which track worker productivity.

Benefits to the environment can be measured by several factors. First, pollutant monitoring prior to the inception of the BMP plan may show significant quantities of pollutants and or wastes that are minimized or eliminated after plan implementation. Discharge monitoring report records may show reductions in the quantity or variability of pollutants in the discharges. In addition, the reductions in volumes of and/or hazards posed by solid waste generation and air emissions may demonstrate the success of the BMP plan.

Other derived environmental benefits may include reduced releases to the environment resulting from spills, volatilization, and losses to storm water runoff. These benefits may be measured through reductions in the number and severity of releases and of lessened losses of materials.

Reduced expenditures are the bottom line in substantiating the need for the BMP plan. Cost considerations can be easily tracked through expense records including chemicals usage, energy usage, water usage, and employee records. The development of production records on product per unit cost before and after BMP plan implementation may show a significant drop, thereby demonstrating the effectiveness of the plan.

2.3.3.2 Plan Reevaluation

The operations at an industrial facility are expected to be dynamic and therefore subject to periodic change. As such, the BMP plan can not remain effective without modifications to reflect facility changes. At a minimum, the BMP plan should be revisited annually to ensure that it fulfills its stated objectives and remains applicable. This time-dated approach allows for the consideration of new perspectives gained through the implementation of the BMP plan, as well as the reflection of new directives, emerging technologies, and other such factors. However, plan revisions should not be limited to periodic alterations. In some cases, it may be appropriate to evaluate the plan due to changed conditions such as the following:

- Restructuring of facility management
- Substantial growth
- Significant changes in the nature or quantity of pollutants discharged
- Process or treatment modifications
- New permit requirements
- New legislation related to BMPs
- Releases to the environment.

Many changes at a facility may warrant modifications to the BMP plan. Growth may require more frequent employee training or a redesign of the good housekeeping program to ensure the site is maintained in a clean and orderly fashion. An evaluation of or modifications to existing process, treatment, and chemical handling methods may substantiate the need for additional facility-specific BMPs.

Where new permit requirements or legislation focus on a specific pollutant, process, or industrial technology, it may be appropriate to consider establishing additional controls. These permit requirements or legislative changes do not necessarily have to be directly related to environmental issues. For example, new OSHA standards may result in modification of the BMP plan to include procedures that address the protection of worker health and safety.

If there has been a spill or other unexpected chemical release, the reasons for the release and corrective actions taken should be investigated. This investigation should include evaluation of all control programs including good housekeeping, PM, inspections, security, employee training, and recordkeeping and reporting. Additionally, facility-specific BMPs should be evaluated at that time to determine their effectiveness.

Ultimately, the BMP plan reevaluation may pinpoint areas of the facility not addressed by the plan, or activities that would benefit from further development of facility-specific BMPs or revision of the general programs contained in the BMP plan. It is useful to bear in mind that as the

BMP plan improves, costs can continue to be minimized as a result of reduced waste generation, less hazardous or toxic materials use, and prevented environmental releases.

3. INDUSTRY-SPECIFIC BEST MANAGEMENT PRACTICES

3.1 PURPOSE OF THIS CHAPTER

Chapter 2 discussed the planning, development, and implementation of a best management practice (BMP) plan, including the scope of the BMP applicability and the components of an effective plan. The intent of this chapter is to illustrate through examples how plan components are implemented in various industries. Examples of BMPs are provided on an industry-specific basis. Exhibit 3-1 presents three basic steps to follow in selecting from among the BMPs provided as examples.

EXHIBIT 3-1. BMP SELECTION PROCESS

1. ***Review the industry profiles to determine the industrial processes that apply.*** Associated with each process are examples of BMP applications. These examples are based on commonly implemented pollution prevention practices, actual case studies, and demonstrations. While this chapter identifies BMPs for a specific industry category, some of the information may also be transferable to other types of industries.

2. ***Evaluate whether the BMP would help to achieve the environmental objectives of the industry.*** Objectives may include reducing discharges of a particular chemical, reducing losses of raw material to the environment, reusing/reprocessing a process solution or material, or minimizing employees' exposure to pollution.

3. ***Consult the references of the document from which the example was obtained.*** Tables summarizing BMPs refer to the source of the information. Generally, the documents cited were obtained either through the Pollution Prevention Information Clearinghouse (PPIC) or the National Technical Information System (NTIS). A supplementary appendix containing a bibliography describes sources further.

This chapter gives the reader specific examples of effective best management and pollution prevention practices, as well as instances where facilities have successfully implemented such practices. The results, including pollution reductions and cost savings, are highlighted for several industries. Most of the information on effective best management and pollution prevention practices is provided in table form. Each listing in the table contains the following:

- ***The BMP:*** The approach taken by the industry to eliminate or reduce a targeted waste.
- ***Targeted processes:*** The industrial activities that generate a targeted waste or releases the waste into the environment.
- ***Targeted wastes:*** Product(s) or byproduct(s) of industrial activity that pose a threat to human health and/or the environment. The targeted waste(s) will be described in the most specific terms possible (i.e., chemical name, elemental components).
- ***Benefits to water:*** Qualitative and qualitative descriptions showing reductions in the water usage or pollutants sent to water media.
- ***Benefits to other media:*** Quantitative and qualitative descriptions showing reductions in pollutants sent to other media.
- ***Incentives:*** Other positive results of the BMP such as financial savings and improvements in safety.
- ***Data sources and page numbers:*** Citations to the bibliography contained in Appendix D that list the document title, the organization that created the document, and the date.

3.2 INDUSTRY CATEGORY SELECTION

Opportunities for pollution prevention through the implementation of BMPs are available for a multitude of industries. Since all industries cannot be addressed in this chapter, the scope has been narrowed to address industries that discharge the greatest quantities of the 17 priority pollutants targeted by EPA's Pollution Prevention Strategy of 1991 based on data contained in the Toxics Release Inventory data base. EPA acknowledges that other methods are available for targeting industries. However, this method was chosen as the best approach after careful evaluation of the information that various methods would provide. EPA believes that this targeting method identifies opportunities where BMPs may be most effective in preventing the greatest quantity of water pollution by the pollutants of greatest concern.

Although this chapter is limited to BMP identification for nine industry categories, the processes identified may be applicable for control of similar processes or pollutants by industries not discussed in this manual. Users of this manual are encouraged to examine all industry categories' information presented in light of other industries possibly having similar processes or pollutants.

3.3 METAL FINISHING

3.3.1 Industry Profile

The category of metal finishing includes manufacturers that take raw metal stock and subject it to various treatments to produce a product at, or closer to, its finished stage. Manufacturers classified as metal finishers perform similar operations that fall under a variety of standard industrial classification (SIC) codes, including industries in major groups 34 (fabricated metal products), 35 (machinery, except electrical), 36 (electrical and electronic machinery, equipment, and supplies), 37 (transportation equipment), 38 (measuring, analyzing, and controlling instruments: photographic, medical, and optical goods; watches and clocks), and 39 (miscellaneous manufacturing industries).

The processes used to treat raw metal stock and, correspondingly, the wastes produced are the common link among the metal finishing category members. Some of these processes are especially amenable to BMPs; that is, implementation of BMPs is relatively easy and results in a significant reduction in the discharge of pollutants. Listed below are processes common among metal finishers and the targeted pollutants that enter wastewater streams:

- ***Electroplating***: Typical wastes produced include spent process solutions containing copper, nickel, chromium, brass, bronze, zinc, tin, lead, cadmium, iron, aluminum, and compounds formed from these metals.

- ***Electroless plating***: The most common wastes produced are spent process solutions containing copper and nickel.

- ***Coating***: Depending on the coating material that is being applied, wastes of concern include spent process solutions containing hexavalent chromium, and active organic and inorganic solutions.

- ***Etching and chemical milling***: Typical solutions used in etching and milling that ultimately enter the wastestream and are of concern include chromic acid and cupric chloride.

- ***Cleaning***: Various organic and inorganic compounds enter the wastewater stream from cleaning operations.

The source of the targeted pollutants are process solutions and raw materials that enter the wastewater stream primarily through rinsing or cleaning processes. A work piece that is removed from a process or cleaning solution is typically subjected to rinsing directly afterwards and carrying excess process contaminants, referred to as dragout, into the rinse tank. The dragout concentrates pollutants in the rinse tank, which is typically discharged into the sewer system.

Another pathway by which targeted pollutants enter the wastewater stream is through the disposal of spent batch process solutions into the sewer system. Spent solutions consist of aqueous wastes and may contain accumulated solids as well. Spent solutions are typically bled at a controlled rate into the wastewater stream. Other sources of pollutants in wastewater streams include clean-up of spills and washdown of fugitive aerosols from spray operations.

3.3.2 Effective BMPs

Numerous practices have been developed to eliminate or minimize discharges of pollutants from the metal finishing industry. Successful source reduction measures have been implemented to eliminate cyanide plating baths, as well as substitute more toxic solvents with less toxic cleaners. In many cases, cleaning with solvents has been eliminated altogether through the use of water-based cleaning supplemented with detergents, heating, and/or agitation. Other source reduction measures have been implemented to minimize the discharges of toxic materials. For example, drain boards and splash plates have been commonly installed to prevent drips and spills. Additionally, the design of immersion racks or baskets and the positioning of parts on these racks or baskets have also been optimized to prevent trapping of solvents, acids/caustics, or plating baths.

The utilization of recycle and reuse measures have also been commonly used. Many facilities have been able to minimize water use and conserve rinsewaters and plating baths by measures including the following:

- Utilizing a dead rinse resulting in the concentration of plating bath pollutants. This solution may be reused directly or further purified for reuse.

- Conserving waters through countercurrent rinsing techniques.

- Utilizing electrolytic recovery, customized resins, selective membranes, and adsorbents to separate metal impurities from plating baths, acid/caustic dips, and solvent cleaning operations.

These operations and measures not only extend the useful life of solutions, but also prevent or reduce the discharge of pollutants from these operations.

Two industries highlighted in this section have implemented best management practices that resulted in substantial cost savings and pollutant reductions. For example, Emerson Electric implemented a program that resulted in savings of more than $700,000 per year and reductions in solvents, oxygen-demanding pollutants, and metals. Best management practices implemented by a furniture manufacturer in the Netherlands resulted in a reduction in metals discharged from 945 to 37 kilograms per year and a decrease in water use from 330,000 to 20,000 cubic meters per year. A detailed discussion of these programs is provided in the following paragraphs. Exhibit 3-2 provides a summary of other examples of demonstrated BMPs.

Emerson Electric, a manufacturer of power tools, implemented a Waste and Energy Management Program to identify opportunities for pollution prevention. An audit resulted in the following actions:

- Development of an automated electroplating system that reduced process chemical usage by 25 percent, process batch dumps by 20 percent, and wastewater treatment cost by 25 percent.

- Installation of a water-based electrostatic immersion painting system to replace a solvent-based painting system. The water-based system resulted in a waste solvent reduction of more than 95 percent.

- Installation of an ultrafiltration system that recovers 65 lbs per day of waste oil and purifies 2,500 lbs per day of alkaline cleaning solution for reuse, which resulted in a reduction of 5-day biochemical oxygen demand loadings to the treatment system of 370 lbs per month. This avoided the need for installation of additional treatment.

- Installation of an alkaline and detergent and steam degreasing system, which resulted in a reduction in waste solvents by 80 percent.

In addition to the reduction of pollutants, Emerson realized annual costs savings of $642,000 in reduced raw material use, $2,200 in reduced water use, and $52,700 in reduced waste disposal.

A furniture manufacturer in the Netherlands reduced metals in its effluent by switching to cyanide-free baths, allowing for longer drip times, using spray rinsing, reusing water, and implementing a closed cooling system. These best management practices, complemented by the installation of treatment technology, reduced metals in the effluent from 945 to 37 kilograms per year. Water use also decreased from 330,000 to 20,000 cubic meters per year.

3.4 ORGANIC CHEMICALS, PLASTICS, AND SYNTHETIC FIBERS (OCPSF)

3.4.1 Industry Profile

The OCPSF industry manufactures more than 25,000 different organic chemical, plastic, and synthetic fiber products. It includes both those facilities whose primary products are organic chemicals, plastics, and synthetic fibers and the facilities that use or produce these chemicals ancillary to their primary production. OCPSF manufacturers have two types of facilities — those with chemical synthesis as their primary function and those that recover organics, plastics, and synthetic fibers as byproducts of other unrelated manufacturers.

OCPSF manufacturers include SIC code 2821 (plastic materials, synthetic resins, and non-vulcanizable elastomers), SIC code 2822 (cellulosic manmade fibers), SIC 2823 (synthetic organic fibers except cellulosic), SIC code 2824 (cyclic crudes and intermediates, dyes, and organic pigments), and SIC code 2869 (industrial organic chemicals not elsewhere classified). All OCPSF products are derived from the same raw materials (methane, ethane, propene, butane, higher aliphatic compounds, benzene, toluene, and xylene). As a result of the variety and complexity of the processes used and of products manufactured, there is an exceptionally wide variety of pollutants found in the wastewaters of this industry including conventional pollutants, metals, and miscellaneous organics resulting from product and byproduct formation.

Contaminated wastewater generation occurs at a number of points, mainly as direct and indirect contact processes, equipment cooling, equipment cleaning, air pollution control systems, and storm water. Direct contact during manufacturing or processing is found in the use of aqueous reaction media. When water is used as a medium for OCPSF chemical processes, a high-strength process wastewater is produced. After the primary reaction has been completed and the final product has been separated from the water media, some residual product and unwanted byproducts remain. Indirect contact process wastewaters include the recovery of solvents and volatile organics from the chemical reaction kettle. Vacuum jets utilize streams of water used to create a vacuum which draws off volatilized solvents and organics from the reaction kettle into solution. Later, recoverable solvents are typically separated and reused while unwanted volatile organics remain in solution in the vacuum water. This wastewater is discharged as process wastewater. Steam ejector systems are similar to vacuum jets, but steam is used instead of water. The steam is then drawn off and condensed, forming a source of process wastewater. Batch processing may require repeated and extensive equipment clean-up between batches, which is usually accomplished with water. Additionally, water scrubbers on emission control devices, and leaks and spills through the plant which contaminate storm water are two other contributors of potentially high concentration of pollutants to discharges.

3.4.2 Effective BMPs

Due to the individuality of many organic manufacturing operations, pollutant and process-specific source reduction and recycle/reuse measures may only be useful in relation to one facility. However, OCPSF manufacturing facilities are similar in that their organic chemicals comprise their raw materials and final products, and the generation and discharge of significant quantities of pollutants result during cleaning processes. Thus, much of the source reduction and recycle/reuse measures have focussed on reducing pollutants lost to wastewater during clean-up.

Dedicated equipment has been purchased by many batch processors to avoid the loss of valuable materials and to prevent the generation of wastewater during clean-up activities. Additionally, many batch processors have provided for the capture of washdown waters for later recycle/reuse. Alternate cleaning methods such as manual wipe down with squeegees also have been shown to help recover organics products which would otherwise be lost during equipment clean-up.

Many OCPSF facilities have implemented programs that use a variety of techniques to minimize pollutant discharges from their plants, save money, and reduce health risks. For example, a paint manufacturer underwent a waste assessment to identify opportunities for implementing pollution prevention practices. Lab experiments that were conducted to test the feasibility of eliminating clean-up steps determined that each 10 percent decrease of waste volume saved $6,000 per year in disposal costs. In response to this finding, the plant rescheduled production to disperse pigments only before batch formulation, which eliminated the need for intermediate storage, reduced the need for solvents such as methyl ethyl ketone, and allowed for cleaning with a small amount of compatible solvent. A process redesign included the development of a production plan that produces paint from light to dark batches, eliminating intermediate clean-up steps which generate wastewater. Ultimately, savings were realized due to reductions in raw materials use, water use, and waste disposal.

Atlantic Industries' Nutley, New Jersey facility reduced water use from 750,000 to 300,000-400,000 gallons per day and also have lessened discharges of organics and other pollutants. One measure that assisted the facility in achieving these reductions involved the simultaneous increase in process chemical concentration, lowering of reaction temperatures, and adoption of new methods for combining dye components. Ultimately, these and other improvements have reduced the amount of organics and inorganics in the wastewater by 50,000 and 250,000 gallons per year, respectively. Other demonstrated BMPs are summarized in Exhibit 3-3.

3.5 TEXTILE MILLS

3.5.1 Industry Profile

Textile mills are manufacturing facilities that transform fiber into yarn, fabric, or other finished textile products. Those mills that fall under SIC major group 23 (apparel and other textile mill products) use dry processes that normally do not result in wastewater discharges. Some of the mills that fall under SIC major group 22 (textile mill products), however, use wet processing. Characteristics of the major wet manufacturing processes and pollutants in the wastewater discharges that may be targeted for BMPs are listed below:

- ***Raw wool scouring*** is the first treatment performed on wool, in which the wool is washed to remove the impurities peculiar to wool fibers. These impurities are present in great quantities, and include grease, sweat, dirt, feces, vegetable matter, disinfectants, and insecticides. It has been estimated that for every pound of fibers obtained, 1½ pounds of waste impurities are produced, mostly dirt, grit, and grease.

- ***Scouring*** is employed to remove natural and acquired impurities from fibers and fabric. The nature of the scouring operation is highly dependent on the fiber type. For example, cotton fabric is sometimes loaded into a pressure vessel containing a solution of sodium hydroxide, soap, and sodium silicate, after which it is completely rinsed to clean the fibers and remove residual alkali. Synthetics, on the other hand, require only light scouring.

- ***Carbonizing*** removes burrs and other vegetable matter from loose wool or woven wool goods to prevent unequal absorption of dyes. The overall water requirements for the carbonization of wool may be substantial. For example, wool is carbonized using sulfuric acid, then rinsed to remove the acid. The wool is then neutralized using a sodium carbonate solution. A final rinse removes the alkalinity.

- ***Fulling*** gives woven woolen cloth a thick, compact, and substantial feel, finish, and appearance. To accomplish this, the cloth is mechanically worked in fulling machines in the presence of heat, moisture, and sometimes pressure. This allows the fibers to felt together, which causes shrinkage, increases the weight, and obscures the threads of the cloth. Fulling is performed by either alkali or acid methods. Fulling is followed by extensive washing to remove process chemicals and prevent rancidity and wool spoilage.

- ***Desizing*** removes the sizing compounds applied to the yarns and is usually the first wet finishing operation performed on woven fabric. It consists of soaking the fabric in a solution of mineral acid or enzymes and thoroughly washing the fabric.

- ***Mercerizing*** increases the tensile strength, luster, sheen, dye affinity, and abrasion resistance of cotton goods. It is accomplished by impregnating fabric with cold sodium hydroxide solution, an alkali solution that causes swelling of cotton fibers. In many mills, the sodium hydroxide is reclaimed in caustic recovery units and concentrated for reuse.

- ***Bleaching*** is a finishing process used to whiten cotton, wool, and synthetic fibers. In addition, bleaching dissolves sizing, natural pectins and waxes, and small particles of foreign matter. It is primarily accomplished with hydrogen peroxide, although hypochlorite, peracetic acid, chlorine dioxide, sodium perforate, or reducing agents may be used.

- ***Dyeing and printing*** are the most complex of the wet processing operations in textile mills. Many mechanisms and many types of dyes are used in coloring textile fibers. Acid dyes are sodium salts, usually of sulfonic acids or carboxylic acids. Azoic dyes are insoluble pigments anchored within the fiber by padding with a soluble coupling compound, usually naphthol. In addition, common salt and surface active compounds are usually necessary to speed the reaction. Basic dyes are usually hydrochlorides of salts or organic bases and are most effective with acrylic fibers. Direct dyes resemble acid dyes in that they are sodium salts of sulfonic acids and are almost invariably azo compounds. Disperse dyes use several carriers such as acetic acid to color cellulose acetate. Mordant dyes have no natural affinity for textile fibers, but dye well when applied to cellulosic or protein fibers that have been mordanted with a metallic oxide such as chromium. Reactive dyes include many methods and chemicals such as sodium chloride, urea, sodium carbonate, sodium hydroxide, and tri-sodium phosphate. Sulfur dyes are complex organic compounds that are insoluble in water but dissolve in a solution of sodium sulfide to which sodium carbonate has been added. Vat dyes, among the oldest natural coloring matters used, are insoluble in water, but become soluble when treated with reducing agents and used with chemicals such as sodium hydroxide. Final washing of the fabric to remove excess dyes and print paste, along with the cleanup of mixing tanks, applicator equipment, and belts, contributes wastewater associated with the dyeing and printing processes. This cleaning process often involves the use of solvents.

In the production of textile products, pollutants generally enter the wastestream during rinsing or cleaning operations. These pollutants may include acids and alkalis such as those used in scouring, carbonizing, fulling, and mercerizing processes. Solvents used for cleanup are predominant at times. Zinc may present a wastewater problem in yarn spinning and manufacturing.

3.5.2 Effective BMPs

BMPs have been successfully applied in the textile industry and range from wastewater recycling and reuse and chemical substitution to process modification and computerization of controls. American Enka, a yarn and thread mill, and United Piece Dye Works, a textile dye and finishing company, have achieved substantial cost savings, reduced pollutant levels in the wastewater effluent, and met permit effluent limits by the successful implementation of best management and pollution prevention practices. A discussion of their successes is presented below. A summary of other proven BMPs in the textile industry is provided in Exhibit 3-4.

For its rayon yarn manufacturing process, American Enka redesigned and implemented a precipitation system to remove and recycle zinc. The redesigned precipitation system involves a two-stage process where, in the second stage, zinc precipitates onto an existing slurry of zinc hydroxide crystals. Sulfuric acid is then used to convert the zinc hydroxide to zinc sulfate. Zinc sulfate is recycled back to the yarn spinning bath. This two-stage process has achieved an estimated savings of $383,000 per year, and removes zinc from the wastewater and solid wastestreams.

United Piece Dye Works was able to meet its effluent limits for phosphorus by materials substitution in the production process, without any capital expenditure. A detailed evaluation of the production processes, process chemistry, and the chemicals used identified the sources of phosphorus. Process modifications to reduce the use of phosphate chemicals, such as hexametaphosphate, and substitution of chemicals not containing phosphate were made. The use of phosphoric acid was eliminated. The level of phosphorus in the wastewater effluent was reduced from 7.7 to less than 1 mg/l through this pollution prevention practice of source reduction.

3.6 PULP AND PAPER PRODUCTS

3.6.1 Industry Profile

Paper and allied products (SIC major group 26) includes manufacturers of pulp, paper and paperboard, and paper products. The six primary subcategories include pulp mills (SIC group number 261), which manufacture pulp from wood (hardwoods or softwoods) or from other materials such as rags, linters, waste paper, and straw; paper mills (SIC group number 262), which manufacture paper and paper products; paperboard mills (SIC group number 263), which manufacture paperboard and paperboard-related products; companies covered under SIC group number 264, which produce converted paper and paperboard products, such as envelopes, non-textile bags, die-cut paper, pressed and molded pulp goods, sanitary paper products, stationary and tablets; manufacturers of paperboard containers and boxes (SIC group number 265), which produce folding paperboard boxes, corrugated and solid fiber boxes, sanitary food containers, fiber cans, tubes, drums and similar products; and companies covered by SIC group number 266, which includes manufacturers of building paper and board from wood pulp and other fibrous materials.

The production of pulp, paper, and paperboard involves four major processes: raw material preparation, pulping and recovery, bleaching, and papermaking. A discussion of each process and its associated wastes is provided below.

- ***Raw material preparation*** includes log washing, bark removal, and chipping and screening processes. These processes can require large volumes of water, but the use of dry bark removal techniques or the recycle of washwater or water used in wet barking operations reduces water consumption.

- ***Pulping and recovery*** reduces raw material into pulp suitable for further processing. Pulp production results in relatively large quantities of wastewater and wastewater pollutants. The wood entering the pulping process consists of cellulose fibers, lignin, semi-cellulose, and other compounds. Lignin, a complex polymer that binds and strengthens wood fibers, is believed to contain dioxin precursors. Pulping processes vary from basic mechanical action, such as groundwood pulping, to complex chemical digesting sequences, such as in the alkaline (soda or kraft), sulfite, or semi-chemical processes.

  Mechanical pulping does not involve use of chemicals; little or none of the wood material is dissolved. Thus, softwoods, which are easier to tear and grind, are typically used in this pulping process. The resultant pulps are generally used in manufacturing newsprint, catalogues, and toweling.

  Chemical pulping removes lignin to enhance fiber flexibility, resulting in a stronger paper product but lower fiber yields (40 to 50 percent). Sulfite pulp may be blended with mechanical pulps as a strengthener and is commonly used in production of viscose rayon, acetate fibers and films, plastic fillers, and cellophanes. Semi-chemical pulping is often used for newsprint, containers, and computer cards. Kraft pulping accounts for approximately 75 percent of the pulp produced for paper and paperboard due to the number of wood types that can be processed. Also, extracts released during the process such as turpentine, tall oil, and resin can be sold separately as commodity chemicals. Pulping process wastes include pulp rejects, cellulosic fines, white water, and chemical recovery wastes.

- ***Bleaching*** results in the removal of color caused by lignins and resins, or by spent cooking liquor from the pulp left by inefficient washing. Therefore, multi-stage bleaching processes are performed to produce light colored or white products. Conventional bleaching involves five stages wherein chlorine is used as the dominant chemical in a series of alternating acid and alkaline bleaching and washing phases. Dioxins, furans, hexachlorobenzene, and hundreds of organochlorine byproducts (acidic, phenolic, and neutral compounds) result from the bleaching process.

- ***Papermaking*** follows pulp preparation processes and encompasses further mixing and blending with non-cellulosic materials occur to create the furnish for paper making. Further preparation steps may include dyeing, sizing, and starching to increase water resistance. The furnish involves a dilute water suspension of pulp, from which a layer of fiber is deposited on a fine screen. Finally, this layer is removed, pressed, dried, and, if desired, coated to form final products. Chemicals used may include titanium, zinc sulfate, lithophone pigments, waxes, starches, sodium silicate, glues, resins, rubber latex, and hydrocarbons. Coating operations typically involve use of chemicals such as polyvinyl chloride, polypropylene, saran lacquer, rubber, acrylic latex, styrene-butadiene latex, polyvinyl acetate, polyvinyl alcohol, and carboxymethyl cellulose.

3.6.2 Effective BMPs

Due to their toxicity and persistence in the environment, chlorinated organics such as dioxins and furans produced by the paper industry are often targeted for pollution prevention. Common practices have included:

- Discontinuing the use of pitch dispersants and defoamers which may contain chlorinated dioxin and chlorinated furan precursor compounds
- Maximizing delignification in the pulping process
- Maximizing brownstock pulp-washing efficiency
- Optimizing bleaching processes through process control monitoring and automation which introduces limited amounts of bleach at specific times
- Utilizing chlorine dioxide or hydrogen peroxide as alternatives to bleach, or in some cases eliminating bleaching altogether.

Although much of the focus has been on source reduction measures targeted at dioxin and dioxin precursor formation, recycle and reuse opportunities are also being utilized. More and more facilities are finding the benefits of recapturing pulp by in-process and final discharge treatment techniques. Facilities are also finding it more economical to utilize dirty water and treatment plant effluent for washing and other miscellaneous processes. In some cases, closed-loop systems can be achieved. Wood slivers and chips screened out during raw material preparation processes can be dewatered in a press and burned in a bark boiler. This process eliminates solid waste while generating power.

Source reduction and recycle methods used by the pulp and paper manufacturing industry have resulted in lessened wastewater and pollutant discharges, which in turn minimizes costs associated with treatment and water usage. Two specific examples of how industry implemented BMPs in the pulp and paper industry are highlighted below. Exhibit 3-5 summarizes BMPs that have been successfully demonstrated in the U.S. and abroad.

One paper mill in England modified the bleaching stage of pulp manufacturing to reduce water usage and reduce the coloration of wastes. This was accomplished by preceding the chlorine, caustic soda, and chlorine dioxide bleaching with oxygen bleaching, which reduced the quantities of reagents and water used in the conventional bleaching process. Rinsewater from the oxygen bleaching can be used to wash cooked pulp, thus reducing the coloration of wastes. The mill reduced water usage by 50 cubic meters per ton of manufactured pulp. Use of caustic soda, chlorine and, chlorine dioxide was also reduced.

A closed-cycle, effluent-free bleached kraft mill in Canada reduced the amount of fresh water needed in the system by 50 percent. Technology innovations include modifications to the bleaching sequence; countercurrent washing to reduce the amount of fresh water needed; reuse of all bleach-plant effluents in the pulp mill; removal of sodium chloride from the white liquor; use of an effluent-free process for generating chlorine dioxide; and installation of spill tanks and other minor changes to facilitate collection and reuse of water throughout the system. The estimated cost to install the system would be $4.5 million (in 1975 dollars) for a 725-ton-per-day mill, where annual savings were reported to be $2.2 million.

3.7 PESTICIDES

3.7.1 Industry Profile

The formulation of pesticides and agricultural chemicals falls into SIC code 2879, and includes companies that formulate and prepare agricultural pest control chemicals or pesticides. In pesticide formulation, highly concentrated organics manufactured elsewhere are converted into pesticide products such as insecticides, herbicides, and fungicides that are ready for use by farmers and gardeners.

There are three types of pesticide formulations: solvent-based, water-based, and solid-based. Solvent-based formulations use a solvent or a solvent-water emulsion as the carrier solution for the active pesticide ingredient. Typical solvents are light aromatics such as xylene, chlorinated organics such as 1,1,1-trichloroethane, and mineral spirits. With water-based formulations, water serves as the carrier for the active pesticide ingredient. Both solvent- and water-based formulations are applied directly in liquid form or propelled as an aerosol. Solid-based pesticide formulations are prepared by blending solid active ingredients with inert solids such as clay or sand. Some dry formulations are prepared by absorbing liquid active ingredients into solid carrier materials. Examples of common solid-based formulations are dusts, wettable powders, granules, treated seed, and bait pellets and cubes.

Pesticide formulating facilities generate wastes during such operations as cleaning of mixing and storage equipment, housekeeping, and laboratory testing for quality assurance. Commonly generated wastewaters include those from equipment clean-up gathered as a result of raw materials left over in containers; pesticide dust and scrubber water from air pollution control equipment; off-specification products; laboratory wastes; spills; waste sands or clays; laundry wastewater; and contaminated storm water runoff.

The significant pollutant parameters in the pesticide industry include organic pollutants, suspended solids, pH, nutrients, the pesticides specific to the product manufactured, metals, phenol, and cyanide. The active ingredients in insecticides include inorganic compounds, organic compounds, chlorinated hydrocarbons, carbamates, and organophosphates. Herbicide formulations include phenoxy, metal organic compounds, triazine, urea, amide, benzoic, and other organic and inorganic compounds. Fungicides utilize organic and inorganic compounds.

3.7.2 Effective BMPs

Many of the circumstances surrounding the individuality of OCPSF manufactured products is shared by pesticide formulators. As such, much of the focus of pesticides formulation source reduction and recycle/reuse measures have been on pollutants released to water during equipment clean-up. Some of the commonly used practices have included:

- Use of dedicated equipment for batch processing to avoid losses of raw materials and products, thereby preventing the generation of wastewater during clean-up activities
- Capturing of washdown waters in tanks for later reuse
- Adoption of cleaning methods such as squeegee wipe down which helps recover pesticide products which would otherwise be lost during equipment washdown.

Other effective source reduction measures have also been practiced in the pesticides formulation industry including the use of dry air pollution control devices and more controlled, efficient batch sequencing.

Dow Chemical reduced chlorinated hydrocarbons in the wastewater effluent by 98 percent and hydrocarbon emissions by 92 percent. They also reduced the volume of packaging wastes by implementing practices such as material substitution, equipment modification, process modification, housekeeping improvements, and periodic assessments of employee performance. Some of the measures implemented by Dow include the following:

- Packaging the pesticide Dursban in 4-ounce water-soluble packages instead of the 2-pound metal cans previously used. This reduces packing waste volume, and eliminates the disposal problem associated with the empty metal cans.
- Shipping the active pesticide ingredient in tank cars instead of 55-gallon drums. This reduces packing wastes and allows tank cars to be rinsed using a solvent present in the ingredient's formulation.
- Adding the drying agent utilizing computers and instream analyzers instead of manual feed and lab analysis. This reduces the wastewater discharged by 37 percent, the chlorinated hydrocarbon in wastewater by 98 percent, and the hydrocarbon emissions to air by 92 percent.
- Implementing other waste minimization measures such as process changes, recycling and reuse programs, and statistical analyses performed on data representing employee performance to pinpoint problem areas and minimize waste.

Other examples of demonstrated BMPs in the pesticides industry are provided in Exhibit 3-6.

3.8 PHARMACEUTICALS

3.8.1 Industry Profile

The pharmaceutical manufacturing industry encompasses the manufacture, extraction, processing, purification, and packaging of chemical materials to be used as medication for humans and animals. Industry products include natural substances extracted from plants and animals, chemically modified natural substances, synthetic organic chemicals, metal-organics, and inorganic materials.

The pharmaceutical industry's SIC codes include 2833 (medicinal chemicals and botanical products), 2834 (pharmaceutical preparations), 2841 (soaps and other detergents, except specialty cleaners), and 2844 (perfumes cosmetics and other toilet preparations). Pharmaceuticals may be manufactured by batch, continuous, and semi-continuous manufacturing operations, but batch production is the most common of these manufacturing techniques. Fermentation, extraction, chemical synthesis, and mixing/compounding/formulating are the processes used in these operations. The pollutants resulting from the manufacturing of pharmaceuticals are described below:

- ***Fermentation:*** Solvents such as methylene chloride, benzene, chloroform, acetone, ethyl acetate, and methanol are most often used in this process. Copper and zinc also are used in fermentation recovery processes. Chemical disinfectants such as compounds containing phenols are used for equipment sterilization.

- ***Biological and natural extraction:*** Most waste from this subcategory is solid waste. Detergents and disinfectants are also normally found in wastewater, as are solvents such as phenol, benzene, chloroform, 1,2-dichloroethane, acetone, and 1,4-dioxane. Ammonia is used to control pH.

- ***Chemical synthesis:*** Benzene and toluene are found in the majority of the process wastestream. Other solvents used include xylene, cyclohexane, pyridine, chloroform, and methylene chloride. Chemical synthesis also generates acids, bases, cyanides, metals, and other pollutants.

- ***Mixing/compounding/formulating:*** Various wastes are generated by these operations and include those pollutants found in the previous operations.

The wastestreams generated during these various processes result from cleaning and sterilizing equipment, chemical spills, rejected products, and the processes themselves. The primary wastewater source is equipment waterwash. Another source is small amounts of non-recyclable waste dust that may be generated during mixing or tableting operations.

3.8.2 Effective BMPs

The pharmaceutical point source category is characterized by a low ratio of finished products to raw materials, especially among drugs produced by fermentation and natural extraction. Disposal and management of the large volumes of raw material waste present both a logistical and a financial burden. Therefore, BMPs that minimize waste generation are important in reducing the release of pollutants into water, air, and soil. By implementing BMPs, many pharmaceutical companies have taken advantage of the dual benefits of reduced waste generation and more cost efficient operations.

As with the OCPSF manufacturing and pesticide formulation industries, pharmaceutical manufacturing is characterized by significant quantities of pollutants and wastewater resulting from equipment clean-up. Many of the source reduction and recycle/reuse discussed in Sections 3.4.2 and 3.7.2 are also applicable to these facilities. Two successfully implemented BMPs are described below; others are presented in Exhibit 3-7.

A pharmaceutical factory producing sophisticated biochemicals, bulk pharmaceutical compounds, and immunochemicals by batch production has considered waste minimization and management a high priority. This plant has enjoyed the benefits of carefully planned BMPs that have improved wastewater discharges and, working conditions and have saved money. BMPs successfully established at this plant include the following:

- Wherever possible, the plant recovers and recycles used solvents. This process saves $292 per batch.
- Butyl acetate vapors are recovered through the dedication of a separate source of vacuum for drying product crystals. Savings from this recovery amount to $26 per batch.

The Merck Rahway, New Jersey, facility has implemented measures which have resulted in the recovery of 229,600 pounds of acetone per year which would normally be discharged. Outside of savings which can be attributed to lessened raw material expenditures, this reuse practice resulted in a reduction in sewer fees of $47,750 per year.

3.9 PRIMARY METALS

3.9.1 Industry Profile

SIC major group 33, primary metal industries, includes facilities involved in smelting and refining of metals from ore, pig, or scrap; rolling, drawing, extruding, and alloying metals; manufacturing castings, nails, spikes, insulated wire, and cable; and production of coke. Major subcategories include blast furnaces, steel works, rolling and finishing mills (SIC group number 331); iron and steel foundries (SIC group number 332); primary and secondary smelters and refiners of nonferrous metals such as copper, lead, zinc, aluminum, tin, and nickel (SIC group numbers 333 and 334); establishments engaged in rolling, drawing, and extruding nonferrous metals (SIC group number 335); and facilities involved in nonferrous castings (SIC group number 336) and related fabricating operations. The main processes common to metal forming operations and the wastes that are typically generated are discussed below:

- ***Sintering:*** This process agglomerates iron bearing materials (generally fines) with iron ore, limestone, and finely divided fuel such as coke breeze. The fine particles consist of mill scale from hot rolling operations and dust generated from basic oxygen furnaces, open hearth furnaces, electric arc furnaces, and blast furnaces. These raw materials are placed on a traveling grate of a sinter machine. The surface of the raw materials is ignited by a gas and burned. As the bed burns, carbon dioxide, cyanides, sulfur compounds, chlorides, fluorides, and oil and grease are released as gas. Sinter may be cooled by air or a water spray at the discharge end of the machine, where it is then crushed, screened, and collected for feeding into blast furnaces. Wastewater results from sinter cooling operations and air scrubbing devices which utilize water.

- ***Iron making:*** Molten iron is produced for steel making in blast furnaces using coke, iron ore, and limestone. Blast furnace operations use water for noncontact cooling of the furnace, stoves, and ancillary facilities and to clean and cool the furnace top gases. Other water, such as floor drains and drip legs, contribute a lesser portion of the process wastewaters.

- ***Steel making:*** Steel is an iron alloy containing less than 1 percent carbon. Raw materials needed to produce steel include hot metal, pig iron, steel scrap, limestone, burned lime, dolomite fluorspar, and iron ores. In steel making operations, the furnace charge is melted and refined by oxidizing certain constituents, particularly carbon in the molten bath, to specified levels. Processes include the open hearth furnace, the electric hearth furnace, the electric arc furnace, and the basic oxygen furnace, all of which generate fumes, smoke, and waste gases. Wastewaters are generated when semi-wet or wet gas collection systems are used to cleanse the furnace off gases. Particulates and toxic metals in the gases constitute the main source of pollutants in process wastewaters.

- ***Casting operations:*** This subcategory includes both ingot casting and continuous casting processes. Casting refers to the procedure of molten metal into a specified shape. Molten metal is distributed into an oscillating, water-cooled mold, where solidification takes place. As the metal solidifies into the mold, the cast product is typically cooled using water, which is subsequently discharged.

- ***Forming operations:*** Forming is achieved by passing metal through cylindrical rollers which apply pressure and reduce the thickness of the metal. Rolling reduces ingots to slabs or blooms. Secondary operations reduce slabs or blooms to billets, plates, shapes, strips, and other forms. Cooling and lubricating compounds are used to protect the rolls, prevent adhesion, and aid in maintaining the desired temperature. Hot rolling generates wastewaters laden with toxic organic compounds, suspended solids, metals, and oil and grease. Cold rolling operations, occurring at temperatures below the recrystallization point of the metal, require more lubrication. The lubricants used in cold rolling include more concentrated oil-water mixtures, mineral oil, kerosene-based lubricants (neat oils), or graphite-based lubricants, which are typically recycled to reduce oil use and pollutant discharges. Subsequent operations may include drawing or extrusion to manufacture tube, wire, or die casting operations. In these operations, similar pollutants are discharged. Contaminated wet scrubber wastewaters may also be generated from extrusion processes but to a lesser degree than in iron and steel making and sintering operations.

- ***Acid pickling:*** Steel products are immersed in heated acid solutions to remove surface scale during pickling operations. This generates wastewater from three sources: (1) rinsewater used to clean the product after immersion in pickling solution; (2) spent pickling solution or liquor; and (3) wastewater from wet fume scrubbers. The first source accounts for the largest volume of wastewater but the second source is very acidic and contains high concentrations of iron and heavy metals.

- ***Alkaline cleaning:*** This process is used when vegetable, mineral, and animal fats and oils must be removed from the metal surface prior to further processing. Large-scale production or situations where a cleaner product is required may use electrolytic cleaning. The alkaline cleaning bath typically contains a solution of water, carbonates, alkaline silicates, phosphates, and sometimes wetting agents to aid cleaning. Alkaline cleaning results in the discharge of wastewaters from the cleaning solution tank, and subsequent rinsing steps. Potential contaminants include dissolved metals, solids, and oils.

3.9.2 Effective BMPs

Primary metals manufacturing operations have experienced source reduction and recycle/reuse benefits similar to those available to metal finishing operations including conserving waters through countercurrent rinsing techniques, and utilizing electrolytic recovery, customized resins, selective membranes, and adsorbents to separate metal impurities from acid/caustic dips and rinsewaters to thereby allow for recycle and reuse.

Some very unique opportunities are also exclusively available to the primary metals industry. For example, the use of dry air control devices and dry cast quench operations have been adopted at some facilities to avoid the generation of contaminated wastewater. Additionally, many facilities are finding markets for byproducts (e.g., sulfides resulting from nonferrous smelting operations can be converted to sulfuric acid and subsequently sold) which avoids the need to discharge these contaminants.

California Steel Industries, Inc., located in Fontana, California reclaimed wastes to increase profits and address water use issues. The facility, a steel mill, is situated in an area that does not have a ready supply of process water. Also, the offsite recycling facility used to dispose of spent process pickle liquor was soon to become unavailable. As a result of these concerns, the company constructed an onsite recycling facility designed to recover ferrous chloride for resale and to reuse water and hydrogen chloride for use in steel processing operations. Environmental benefits include the recovery and resale of 20 to 25 tons per day of ferrous chloride, 3,550 gallons per day of hydrogen chloride, and 13,000 gallons per day of water. In addition, corporate liability was minimized because spent liquor was no longer sent to a disposal facility.

Exhibit 3-8 provides a summary of other effective BMPs for the primary metals industry.

3.10 PETROLEUM REFINING

3.10.1 Industry Profile

The petroleum refining industry uses chemical reactions and physical separation processes to create gasoline, residual fuel oil, jet fuel, heating oils and gases, petrochemicals, and a wide variety of other products from crude petroleum. Businesses classified as petroleum refining facilities are represented by SIC group number 2911.

A petroleum refinery is a complex combination of interdependent operations engaged in separating crude molecular constituents, molecular cracking, molecular rebuilding, and solvent blending and finishing to produce petroleum-derived products. More than 150 separate processes have been identified for the refining of crude petroleum and its products. Each unit of operation may be associated with quite different water usages. The types and quantities of contact wastewater produced are directly related to the nature of the various processes. Some major petroleum refining processes and associated wastewater pollutants are described below.

- ***Crude oil and product storage:*** Crude oil, intermediate, and finished products are stored in tanks of varying sizes to provide adequate supplies for various refining processes. Operating schedules usually permit sufficient detention time for settling of water and suspended solids. Pollutants are mainly in the form of emulsified oil and suspended solids. Wastes are also a result of spills, leaks, and tank cleaning.

- ***Ballast water storage***: Tankers which are used to ship intermediate and final products generally discharge ballast. Ballast waters are contaminated with product materials that are the crude feedstock in use at the refinery, ranging from water soluble alcohol to residual fuels, and brackish water.

- ***Crude desalting:*** Salts are separated from oil using emulsifiers and settling tanks. The wastewater stream from a desalter contains emulsified oil, ammonia, phenol, sulfides, suspended solids, and chlorides. Thermal pollution is also a problem in that the wastewater often exceeds 95° Celsius.

- ***Crude oil fractionation:*** Fractionation serves as the basic refining process for the separation of crude petroleum into intermediate fractions of specified boiling point ranges. Wastes include sour water drawn off from overhead accumulators prior to recirculation, which contains sulfides, ammonia, oil, chlorides, mercaptans, and phenols. Discharge from oil sampling lines also may contribute pollutants to wastewaters.

- ***Catalytic cracking:*** Catalytic cracking breaks heavy fractions such as oils into lower molecular weight fractions. This process produces large volumes of high-octane gasoline stocks, furnace oils, and other middle molecular weight distillates. Pollutants generally come from steam strippers and overhead accumulators on fractioners. Major pollutants resulting from catalytic cracking are oil, phenols, sulfides, cyanides, ammonia, and carbon monoxide.

- ***Solvent refining:*** Solvents are used to extract contaminants from stock. The major pollutants from solvent refining are the solvents themselves. Under ideal conditions the solvents are continually recirculating with no losses to the sewer. Unfortunately, some solvent is always lost through pump seals, flange leaks, and elsewhere. Solvents are mostly lost from the bottom of fractionation towers and include phenol, glycol, and amines.

3.10.2 Effective BMPs

The petroleum refining industry is unique in that its raw materials, wastes, and products are the same. Thus, source reduction measures such as materials substitution for crude oil are not realistic. Some facilities have, however, begun only accepting crude oil which meets certain quality specifications. Other facilities have implemented source reduction measures involving the use of less toxic catalysts and additives.

Recycle and reuse opportunities for the petroleum refining operations are plentiful. Tank bottoms, slop oil, dissolved air flotation float, and American Petroleum Institute separator sludge are commonly recycled to the crude unit and, in some cases, the coker for further onsite processing to recover hydrocarbons. Many other by-products also can be recovered and reused onsite. For example, recovered acids and caustics can be used for wastewater neutralization. Other recovered materials (i.e., spent catalyst) are often shipped offsite for reuse in the paper industry or for further reclamation of precious metals such as vanadium. The proven use of BMPs by one industry is highlighted below. Exhibit 3-9 provides a summary of other effective BMPs.

A large petroleum refining operation installed a Stretford Chemical Recovery Process (SCRP) unit to recover active sulfurs in wastewaters. By recovering these sulfurs, the frequency of solution dumping decreased from every 2 ½ months to once per year, with a reduced disposal volume of 225,000 to 25,000 gallons per year. Thus, releases of sulfur to the wastewater and the need for dumping and offsite disposal of sulfur were also reduced. This helped minimize risks of soil contamination at disposal sites. Ultimately, the savings including $60,000 per year in disposal costs and $120,000 per year in raw materials resulted.

3.11 INORGANIC CHEMICALS

3.11.1 Industry Profile

The inorganic chemicals industry is very large and diversified. Thirty-five major categories of inorganic chemicals known to generate polluted wastewater are listed in Exhibited 3-10. The major industries included manufacturers of alkalies and chlorine (SIC group number 2812), industrial gases (SIC group number 2813), inorganic pigments (SIC group number 2816), and industrial inorganic chemicals not elsewhere classified (SIC group number 2819).

Inorganic chemicals are manufactured for captive or merchant use in four or more steps moving from raw material to final product. Two or more different products may use the same process, but the raw materials used, process sequence, control, recycle potential, handling, and quality control varies among products, as does the quality of wastes.

Plant process wastewaters from the inorganic chemicals industry often contain toxic metals such as mercury, zinc, chromium, lead, arsenic, cadmium, nickel, silver, copper, and cyanide. Very few organic toxic pollutants are found in process wastestreams, and those found tend to be present in low-level concentrations. All of the processes tend to have discharges of acids such as sulfuric, hydrofluoric and hydrochloric. Other substances often present in the wastewater are salts, asbestos, total residual chlorine, and iron.

EXHIBIT 3-10. TYPES OF INORGANIC CHEMICALS

Chlor-Alkali	Hydrochloric Acid	Boric Acid
Hydrofluoric Acid	Nitric Acid	Calcium Carbonate
Titanium Dioxide	Sodium Carbonate	Cuprous Oxide
Aluminum Fluoride	Sodium Metal	Manganese Sulfate
Chrome Pigments	Sodium Silicate	Strong Nitric Acid
Hydrogen Cyanide	Sulfuric Acid	Oxygen and Nitrogen
Sodium Dichromate	Carbon Dioxide	Potassium Iodide
Copper Sulfate	Carbon Monoxide	Sodium Hydrosulfide
Nickel Sulfate	Silver Nitrate	Sodium
Sodium Bisulfite	Ammonium Chloride	Silicofluoride
Sodium Thiosulfate	Sodium Hydrosulfite	Ammonium Hydroxide
Sulfur Dioxide	Hydrogen Peroxide	Barium Carbonate

3.11.2 Effective BMPs

BMPs directed at reducing wastewater consumption have greatly reduced costs in the inorganic chemical manufacturing industry. While treatment technologies have been most often used to reduce the levels of toxic metals and other pollutants in the wastewater, source reduction and recovery/reuse practices have also been employed.

Many of the most effective pollution prevention mechanisms in the inorganic chemicals manufacturing industry have employed the reclamation of byproducts, which had previously been considered wastes, and the development of saleable products. Additionally, as with the OCPSF manufacturing industry, the inorganics manufacturing industry has found success in substituting less toxic catalysts and in exhibiting better reaction controls, thus minimizing or eliminating excess toxic and hazardous wastes.

One industry's success in implementing BMP programs is discussed below. Other effective BMPs in the inorganic chemicals industry are summarized in Exhibit 3-11.

A European manufacturer of industrial inorganic chemicals instituted recycling for desalination water produced during the production of hydrazine. This resulted in a reduction in wastewater effluent generation by more than 90 percent. In addition, chemicals such as hydrogen peroxide and ammonia are recovered and recycled and mineral residues are recovered and sold to a cement works. After the introduction of the BMPs, process water usage was reduced by 90 percent, energy used was reduced by 60 percent. Ultimately, this contributed to operating costs being 40 percent lower.

EXHIBIT 3-2. SUMMARY OF BMPs UTILIZED IN THE METAL FINISHING INDUSTRY

BMP	Targeted Process(es)	Targeted Waste(s)	Benefits to Water	Benefits to Other Media	Other Incentives	Data Source
Source Reduction: Frequent inspection of plating rack and tank liners for loose insulation	Plating line	Ferrocyanid	Maintenance of tank linings prevents the formation of ferrocyanides, which are difficult to treat in wastewater			F1
Source Reduction: Separation of cyanide wastes from nickel or iron wastes	Plating line	Cyanide, cyanide compounds, nickel, nickel compounds	Prevents the formation of cyanide complexes, which are difficult to treat			F1
Recycle/Reuse: Installation of an ion exchange recovery system to recover nickel	Plating line	Nickel	Reduces nickel concentrations in the wastewater by recovering nickel sulfate solution and returning it to nickel plating tanks		Reduces need for new plating chemicals. Wastewater treatment and disposal costs were reduced by $2,600 per year.	F2
Source Reduction: Modification to include an automated electroplating system	Plating line	Typical process solution pollutants	Reduces process chemicals in wastewater		Reduces raw material costs by $8,000. Reduces wastewater treatment costs by 25 percent.	F2
Recycle/Reuse: Installation of an ultrafiltration system for recovery of alkaline degreaser and oil for later reuse	Wastewater treatment	Waste oil and alkaline cleaning solution	The ultrafiltration removes alkaline degreaser and oil, thereby reducing the BOD loading to the wastewater treatment system by 370 lbs per month		Allows for the recovery of oil at a savings of $8,000 per year and of alkaline degreaser at a savings of $3,000 per year	F2

EXHIBIT 3-2. SUMMARY OF BMPs UTILIZED IN THE METAL FINISHING INDUSTRY (Continued)

BMP	Targeted Process(es)	Targeted Waste(s)	Benefits to Water	Benefits to Other Media	Other Incentives	Data Source
Source Reduction and Recycle/Reuse: Process modification to add nickel dead rinse, increase drain time, install spray rinses, implement copper and chromium rinse recycling, and lower plating bath concentration while increasing plating bath temperature	Plating line	Copper, chromium, and nickel	Nickel dead rinse helps concentrate nickel for easy recovery, as well as removing it from the wastewater. Other measures reduce the introduction of pollutants to the wastewater.	Recycling copper and chromium rinses eliminates plating sludge and solid waste disposal	Reduces waste disposal costs	F3
Recycle/Reuse: Installation of a closed loop system with filtration, ion exchange resins, and electrolytic recovery	Plating line	Typical process solution pollutants and precious metals	The filtration, ion exchange resins, and electrolytic recovery purify the wastewater, enabling the water to be recycled. The quantity of wastewater produced is reduced.	The filters, resins, and cathodes are sent to refiners for reclamation. No wastes are lost from the process. Thirty-six tons per year of metal hydroxide sludge are eliminated.	The capital cost of the system was $220,000. The facility annually saves $45,000 from feedstock reduction and $23,000 from reduction in sludge production.	F4
Source Reduction: Replacement of hexavalent chromium baths with trivalent chromium baths	Plating line	Chromium	Generates trivalent chromium, much less toxic and easier to treat than hexavalent chromium. The toxicity of the wastewater is reduced.		Wastewater treatment costs are reduced	F5

EXHIBIT 3-2. SUMMARY OF BMPs UTILIZED IN THE METAL FINISHING INDUSTRY (Continued)

BMP	Targeted Process(es)	Targeted Waste(s)	Benefits to Water	Benefits to Other Media	Other Incentives	Data Source
Recycle/Reuse: Installation of an ion exchanger and an evaporator unit	Plating line	Chromium, cyanide, and typical process solution pollutants	The rinsewater is pumped to an ion exchanger to prevent the buildup of metal impurities. The rinsewater then enters an evaporator where it is concentrated before returning to the plating bath for reuse with chromium. Upon return to the plating bath, the rinsewater is chemically treated to remove impurities. Wastewater discharges are reduced and potentially eliminated.		Saves $14,300 per year by recycling chromium and reducing the volume of sludge for disposal	F6
Source Reduction: Replacement of degreasing solvents by installing an alkaline detergent and steam degreasing system	Rinsing	Solvents	Reduces toxic organic chemical loadings to the wastewater treatment system		Reduces quantity of solvents used by 80 percent. New system cleans six times more effectively, thereby reducing air emissions.	F7
Source Reduction: Installation of a drain board between the plating bath and rinse bath or another process bath	Plating line	Nickel, chromium, and other typical process solution pollutants	The drain board routes spillage back to the rinse line, thereby requiring less rinsewater and also reducing the pollutant load entering wastewater by spills during drainage. In one case, the plating bath constituents entering the wastewater stream were reduced from 7 to 1 lbs per day.		By routing the drained plating solution back to the plating bath, process solution is conserved at minimal cost	F7

EXHIBIT 3-2. SUMMARY OF BMPs UTILIZED IN THE METAL FINISHING INDUSTRY (Continued)

BMP	Targeted Process(es)	Targeted Waste(s)	Benefits to Water	Benefits to Other Media	Other Incentives	Data Source
Source Reduction: Positioning of the work piece so that (1) the surface is as close to vertical as possible; (2) the longer dimension of the work piece is horizontal; and (3) the lower edge of the work piece is slightly tilted so that drips occur from a corner rather than an edge	Plating line and rinsing	Typical process solution pollutants	By properly positioning the work piece, the amount of dragout from the system is decreased. This in turn results in requirements for less rinsewater and a reduced pollutant load going in the wastestream.			F7
Source Reduction: Changing location of incoming water to the end of the tank furthest from where work is introduced; installation of flow control valve; and installation of air agitation line diagonally across the tank	Rinsing	Typical process solution pollutants	Reduces the amount of rinsewater discharged		Reduces the costs of water usage	F7
Source Reduction: Installation of a three-tank counter-current rinsing system which exposes the work piece to the most contaminated rinsewater tank first; fresh water flows from the least contaminated tank to the most contaminated tank	Rinsing	Typical process solution pollutants	Reduces rinsewater discharges dramatically. The effluent from the most contaminated tank can be routed to the plating bath, reducing the wastewater effluent.		Reduces the costs of water usage	F7

EXHIBIT 3-2. SUMMARY OF BMPs UTILIZED IN THE METAL FINISHING INDUSTRY (Continued)

BMP	Targeted Process(es)	Targeted Waste(s)	Benefits to Water	Benefits to Other Media	Other Incentives	Data Source
Source Reduction and Recycle/Reuse: Installation of a static rinse tank immediately following plating bath	Planting line	Typical process solution pollutants	Reuses plating bath constituents as makeup after concentration builds up in the tank. Decreases the pollutant loading to the wastestream by reusing the dragout. Requires less water since static rinse removes most contaminants.		Conserves process solutions to reduce expenditures	F7
Source Reduction: Removal of the work piece from the bath as slowly and as smoothly as possible, allowing ample time for the piece to drain over the plating bath	Plating line	Typical process solution pollutants	The amount of dragout from the system is decreased. Requires less rinsewater. Reduces pollutant load to the wastestream.		Conserves process solutions to reduce expenditures	F7
Source Reduction: Institution of changes including (1) conversion of countercurrent rinses into one static rinse and one large continuous rinse; (2) installation of spray rinse and drip pans to allow longer drip time; (3) adjustment of nozzle size and duration of rinse time; and (4) installation of electropurification unit in the chromium plating tank	Plating line and rinsing	Primarily chromium	May reduce the loading of chromium in the wastewater by up to 85 percent		Realized annual savings of $7,000 from the recovery of chromium for reuse and from lower costs of treatment	F8
Recycle/Reuse: Installation of ion exchange system to treat wastewater, electrolytic recovery unit to reclaim metals, and chemical precipitation system to generate metal hydroxide	Plating line	Chromium, copper, cyanide, and nickel	Reduces rinsewater volume and metals concentrations such that wastewater meets regulations	Eliminates sludge production	Recovers metals for reuse and eliminates cost of sludge disposal	F8

EXHIBIT 3-2. SUMMARY OF BMPs UTILIZED IN THE METAL FINISHING INDUSTRY (Continued)

BMP	Targeted Process(es)	Targeted Waste(s)	Benefits to Water	Benefits to Other Media	Other Incentives	Data Source
Recycle/Reuse: Installation of a closed loop system with an electrolytic recovery cell	Plating line	Copper	The electrolytic cell recovers copper from the rinsewater, enabling the water to be reused. The quantity of wastewater produced is reduced.	The volume of sludge is reduced by 2,860 lbs per year	The copper is periodically removed from the cell and resold as scrap, resulting in a cost benefit of $2,000 per year. Sludge disposal costs are reduced by $4,000 per year.	F9
Recycle/Reuse: Switch from single pass to closed loop system allowing for completion of oxidation per reduction treatment prior to centrifugation	Plating line	Typical process solution pollutants	The recycling of the rinsewater reduces the volume of water used and reduces wastewater generated by 40 percent	Reduces volume of sludge produced and the amount of chemicals required for treatment	Capital costs of $210,000 were recovered in 36 months due to decreased disposal costs. Realized total savings of $58,460 per year.	F9
Source Reduction and Recycle/Reuse: Installation of an atmospheric evaporator combined with an electrolytic cell	Plating line	Chromic acid and copper	The electrolytic cell recovers dissolved copper in the dragout. The atmospheric evaporator recovers the chromic acid dragout. This reduces the copper and chromium pollutant levels in the wastestream.	The evaporator eliminates atmospheric discharges. Reduces the amount of sludge generated.	Recovers 92 percent of the copper and 95 percent of the chromium, thereby reducing expenditures on new raw materials	F9
Source Reduction and Recycle/Reuse: Installation of an evaporation system on an electroplating bath	Plating line	Nickel chloride, nickel sulfate, and boric acid	The evaporation of excess rinsewater allows all of the wastewater to be returned to the plating bath as make-up water and eliminates the wastewater discharge from the plating line	The production of wastes requiring disposal in a landfill has decreased by 50 percent	Capital costs for the system installation were recovered in 7 months	F10

EXHIBIT 3-2. SUMMARY OF BMPs UTILIZED IN THE METAL FINISHING INDUSTRY (Continued)

BMP	Targeted Process(es)	Targeted Waste(s)	Benefits to Water	Benefits to Other Media	Other Incentives	Data Source
Source Reduction and Recycle/Reuse: Installation of air per water sprays for rinsing parts; replacement of overflow rinse tanks with static rinse tanks; pumping rinsewater to heated tanks for evaporation	Plating line	Chromium compounds and cadmium compounds	The spray rinse requires less rinsewater and the evaporation of the excess rinsewater allows the rinsewater to be recycled, eliminating all wastewater discharge		Realized financial benefits from reduced water usage, reduced wastewater treatment costs.	F10
Source Reduction and Recycle/Reuse: Installation of spray rinse and spray rinse tank, and drip bar in still rinse tank; conversion of countercurrent rinses to one still rinse tank and one large continuous rinse; installation of electropurification unit in plating tank	Plating line	Chromium compounds	Initial spray rinse of the work piece over the plating tank reduces dragout in the wastestream because the rinsewater returns to the tank. The electropurification unit enables the dragout to be reused by removing contaminants. The new rinse system reduces water flow from 1.2 gallons per minute (gpm) to 1.0 gpm.	The reduction of dragout results in a reduction in sludge production	The system reduces the need for raw materials and treatment chemicals. The initial capital cost of the system was $2,900. Chromium and treatment chemical expenditures were reduced by $7,000.	F11
Source Reduction and Recycle/Reuse: Installation of activated carbon filtration system	Plating line	Typical process solution pollutants	Removes the impurities in spent plating baths using activated carbon filtration. This enables the baths to be reused, eliminating their disposal, and reducing the volume of plating line wastewater by 10,800 gallons per year.		Decreased volume of plating baths disposed and reduced raw material requirements. Capital costs of $9,192 were recovered in 3 months.	F12
Source Reduction and Recycle/Reuse: Installation of a closed loop plating system with filtration of plating waters	Plating line	Trichloroethane and other process solution pollutants	Reduces wastewater production by 95 percent	Decreases the amount of trichloroethane use by 50 percent	Reduces virgin solvent requirements and recovers metals, thereby reducing expenditures	F13

EXHIBIT 3-2. SUMMARY OF BMPs UTILIZED IN THE METAL FINISHING INDUSTRY (Continued)

BMP	Targeted Process(es)	Targeted Waste(s)	Benefits to Water	Benefits to Other Media	Other Incentives	Data Source
Source Reduction and Recycle/Reuse: Installation of a system that reduces pressure to vaporize water at cooler temperatures and condenses vapors in another container for recycling. Soluble pollutants are concentrated and precipitated out of solution.	Plating line	Typical process solution pollutants	Reduces the quantity of wastewater produced by recovering pure water			F14
Source Reduction and Recycle/Reuse: Process modification to include countercurrent rinsing in combination with reverse osmosis and ozone treatment system	Plating line	Chromium and cyanide compounds	Allows for recycling of wastewater. This enabled the facility to eliminate the system's wastewater discharge and reduce water consumption by 17,600 gallons per day.			F15
Recycle/Reuse: Installation of an advanced reverse osmosis system for recovery of nickel plating bath solution and rinsewater	Plating line and rinsing	Nickel	Allows for recycling of rinsewater and removal of nickel from wastewater discharge		Generates nickel concentrate at concentrations of 40 to 50 percent nickel which are readily useable as plating bath solution makeup. Provides a payback period of 4.4 years.	F16
Source Reduction: Replacement of trichloroethane and methanol with terpene-based cleaner	Degreasing	Trichloroethane and methanol	Eliminates the discharge of more toxic trichloroethane and methanol pollutants		Analyses showed that the dilute solution adequately removed contaminants and no residual was detected on the parts. The quality of the coating bond was slightly better for the dilute cleaner. Resulted in a payback period of 4.5 months.	F16

EXHIBIT 3-2. SUMMARY OF BMPs UTILIZED IN THE METAL FINISHING INDUSTRY (Continued)

BMP	Targeted Process(es)	Targeted Waste(s)	Benefits to Water	Benefits to Other Media	Other Incentives	Data Source
Source Reduction: Use of a thinner metal foil on the plastic board in manufacturing printed circuit boards to reduce the amounts of etching solution used	Manufacture of printed circuit board	Spent bath solution (including cyanides, hexavalent chromium, copper, nickel, zinc, and cadmium)	Allows for the use of more dilute solutions for equivalent plating. This reduces the amount of spent bath waste and the corresponding concentrations of bath chemicals in the wastewater discharge. Rinsewater quantities were also reduced.		Requires less etching solution and reduced the hazardous nature of the spent bath	F17
Source Reduction: Replacement of chlorofluorocarbons with Simple Green, an alkaline-based detergent	Degreasing	Chlorofluoro-carbons	Eliminates the discharge of chlorofluorocarbons by substituting a non-toxic biodegradable detergent	Eliminates chloroflourocarbon emissions which damage the ozone layer	Since Simple Green is biodegradable, reduces waste management costs	F18
Recycle/Reuse: Removal of arsenite or arsenate salts from etching acid solution	Etching	Arsenic	Allows for the reuse of the acid and eliminates arsenic from the wastewater discharge		Reusing the acid would result in cost savings	F19
Recycle/Reuse: Replacement of arsenic with an alternate means of providing oxidation protection	Preparation of surfaces	Arsenic	Eliminates arsenic from the wastewater discharge			F19
Source Reduction: Installation of feed-back and feed-forward control loops to determine when process bath, reaction mixture, or rinsewater discharge is necessary	Process bath, reaction mixture, and rinsewater discharging	Process bath, reaction mixture, and rinsewater	Installing feed-back and feed-forward control loops allows for the accurate determination of when a process bath, reaction mixture, or rinsewater has reached its safe loading, thus decreasing unnecessary dumping and reducing waste		Results in cost savings from reduced expenditures for process baths, reaction mixtures, and rinsewater	F19

EXHIBIT 3-2. SUMMARY OF BMPs UTILIZED IN THE METAL FINISHING INDUSTRY (Continued)

BMP	Targeted Process(es)	Targeted Waste(s)	Benefits to Water	Benefits to Other Media	Other Incentives	Data Source
***Recycle/Reuse*:** Improvement of the recoverability and reuse capability of cyanide-containing plating baths	Cyanide recovery	Cyanide	Allows for the recovery and reuse of cyanide and minimizes discharges of cyanide		Provides for an extension of the bath life and a reduction in the frequency that baths must be discarded	F19
Source Reduction: Use of carbon dioxide blasting in lieu of water- or solvent-based cleaners to remove paint	Paint removal	All wastewater	Eliminates the generation of discharges since carbon dioxide blasting is a dry process	Air emissions contain only carbon dioxide, no organic solvents	Reduces operating costs since masking is not required. Reduces aircraft down times since many blast nozzles are used at one time.	F20
Source Reduction: Replacement of freon degreasing and drying solvents with biodegradable detergent solution	Degreasing and drying	Freon	By eliminating freon from the degreasing and drying processes, freon was removed from the wastewater discharge	By eliminating the use of freon, the risk of ozone depletion was minimized	Improved worker safety and cost savings result from the use of the biodegradable detergent solution	F20

EXHIBIT 3-3. SUMMARY OF BMPs UTILIZED IN THE OCPSF MANUFACTURING INDUSTRY

BMP	Targeted Process(es)	Targeted Waste(s)	Benefits to Water	Benefits to Other Media	Other Incentives	Data Source
Source Reduction: Collection of mineral spirits used for equipment clean-up and use of these waste spirits in subsequent batch formulations	Equipment cleaning	Solvents	Reduces wastewater generated during mixing of paints			C1
Source Reduction: Use of a new separation technology which eliminates pollutants from discharge waters	Caprolactam manufactured from toluene	Ammonium sulfate byproduct	The new separation step eliminates the use of ammonia therefore no ammonium sulfate forms. Also eliminates the aqueous wastestream.			C2
Recycle/Reuse: Sequencing of washwaters from small vessels to large, then to a filter press. Concentrated wastewater from the filter press is then reused in the production batch.	Equipment cleaning	Typical process constituents	Sequencing washwater conserves the amount of water used and reusing washwater reduces wastewater discharges		Decreases the costs to wastewater treatment and water usage	C3
Source Reduction: Frequently changing filter press alignment and filter cloth for leaks	Production	Typical process constituents	Preventing leaks will keep dye from entering the wastestream, thus reducing the pollutant loads to the discharge		Less final product is lost	C3
Recycle/Reuse: Reuse of filter washwaters for cleaning equipment and floors	Equipment cleaning	Water	Reusing filter washwaters lessens the amount of wastewater generated		Decrease costs associated with water usage	C3
Source Reduction: Installation of high pressure sprays for filter press and centrifuge clean-up	Equipment cleaning	Water	High pressure sprays expedite the removal of filter cake from the filter equipment. Garden hose type spay nozzles use 10-100 times the amount of water as high pressure spays.		Decrease costs associated with water usage	C3

EXHIBIT 3-3. SUMMARY OF BMPs UTILIZED IN THE OCPSF MANUFACTURING INDUSTRY (Continued)

BMP	Targeted Process(es)	Targeted Waste(s)	Benefits to Water	Benefits to Other Media	Other Incentives	Data Source
Source Reduction: Use of phosphates as corrosion inhibitor instead of chromates	Cooling	Chromium and zinc	Eliminated chromium and zinc from the cooling water wastestream, resulting in regulatory compliance			C4
Source Reduction: Implementation of a stewardship program to reduce fluctuations of organic contaminants in wastewater	Solvent and chemical additives manufacturing	Typical process constituents	Establishes sampling stations to monitor concentrations of organic chemicals. Fluctuations were quickly identified and remedied. This program led to a 75 percent reduction of organic wastes entering the wastestream.			C5
Recycle/Reuse: Use of a filter rinse to recover phenolic resins	Urea and phenolic resins manufacturing	Large phenolic resin particles	Phenolic resins are rinsed into large tanks and recycled into the process as raw material. These resins were previously rinsed into the wastewater treatment system.			C5
Recycle/Reuse: Use of a rinse process modification reduces the volume of phenolic waste	Urea and phenolic resins manufacturing	Phenolic resin waste	New rinse procedure utilizing phenolic resin reactor vessels reduces volume of rinsewater by 95 percent, resulting in a more concentrated solution that is recycled into the process line as raw material.			C5
Source Reduction: Use of process modification to eliminate the use of mercury	Aminoanthra-quinone manufacturing	Mercury	A novel chemical pathway was developed to circumvent the sulfonation step in producing aminoanthraquinone, thus eliminating the need for mercury as a catalyst. This eliminates the discharge of 58 pounds per year of mercury to wastewater.	This process eliminates the discharge of 10 pounds of mercury emissions to the air per year and 325 pounds of mercury to solid waste per year		C6
Source Reduction: Change of a production process	Dye manufacturing	Chromium	New process uses chromium far more efficiently, resulting in 25 percent less chromium entering the wastewater stream		Reduces long term liability associated with chromium wastes	C6

EXHIBIT 3-3. SUMMARY OF BMPs UTILIZED IN THE OCPSF MANUFACTURING INDUSTRY (Continued)

BMP	Targeted Process(es)	Targeted Waste(s)	Benefits to Water	Benefits to Other Media	Other Incentives	Data Source
Source Reduction: Modification of production process	Sulfonation reactons	Sulfuric acid	Lab and plant scale experimentation identified minimum amounts of acid needed to ensure completion of its major sulfonation reactions, thereby reducing sulfuric acid waste discharge to wastewater by 10 to 40 percent			C6
Recycle/Reuse: Use of an incinerator, installation of expanded storage capacity, and installation of concentration equipment	Air pollution control and manufacturing processes	Hydrochloric acid	Led to the use of hydrochloric acid onsite as a raw material. Excess acid is sold as a commercial product. These developments eliminated the discharge of hydrochloric acid to wastewaters.		Allowed the company to decide not to renew permits to discharge hydrochloric acid. This reduction option also offered economic advantages in selling what was once part of a wastestream.	C6
Source Reduction: Use of countercurrent washing in phenolic resins manufacturing to reduce washwater use	Rinsing	Water	When the product resin needs to be washed several times, countercurrent washing generates low quantities of wastewater			C7
Recycle/Reuse: Reuse of resin washwater	Rinsing	Water	By reusing resin washwater from the fist rinse, total water usage decreased		Allows for recovery of phenol from the washwater accumulated from the first rinse	C7
Source Reduction: Elimination of the use of toxic catalysts in manufacturing dyes and pigments	Dye manufacturing	Toxic metals	Eliminates toxic catalyst residuals from the wastewater discharge			C7
Source Reduction: Precise measurement of chemicals based on the stoichiometric formulation of process batches	Dye manufacturing	Dyes and chemicals	Eliminates excess chemicals which results in the introduction of fewer toxins to the wastewater discharge		Provides savings in the costs of handling, storage, and transport of dyes and chemicals	C8

EXHIBIT 3-4. SUMMARY OF BMPs UTILIZED IN THE TEXTILES MANUFACTURING INDUSTRY

BMP	Targeted Process(es)	Targeted Waste(s)	Benefits to Water	Benefits to Other Media	Other Incentives	Data Source
Source Reduction and Recycle/Reuse: Installation of a heat exchange system to capture heat from spent dye solutions to preheat subsequent dye operations	Dyeing	Heat	Lowers temperature of effluent entering the municipal sewer system from 130°F to 70°F		Conserves energy	T1
Recycle/Reuse: Installment of reverse osmosis system to treat rinsewater	Rinsing	Mineral oil lubricant carried away by a continuous rinse process	Reduces BOD concentration arising from mineral oil in wastewater discharge		Reduces costs of replacing mineral oil and reduces processing costs	T2
Source Reduction: Installation of a control system for the dressing of paste used to print textiles	Pigmentary printing of cloth	White spirit, white spirit fumes, and reusable tank sediments	Better process control results in an 87 percent reduction in white spirit in the washing water effluent	A 70 percent reduction was observed in white spirit fumes generated during the drying stage		T3
Zinc Recovery and Recycling: Use of a solution of D.E.H.P.A. (10 percent) and solvesso (90 percent)	Rayon production	Zinc	Reduces zinc		Reduced costs associated with purchasing zinc	T3
Recycle/Reuse: Reuse of the dye bath	Dyeing	Dye and other specialty chemicals	Reduces waste in effluent		Conserves dye and other specialty chemicals. Conserves energy by avoiding the reheating of the dye bath.	T4
Source Reduction and Recycle/Reuse: 100 percent wastewater reuse	Synthetic textile manufacture	Typical process pollutants	Eliminates wastewater discharge		Reduces consumption of fresh water (which must be purchased)	T5

EXHIBIT 3-4. SUMMARY OF BMPs UTILIZED IN THE TEXTILES MANUFACTURING INDUSTRY (Continued)

BMP	Targeted Process(es)	Targeted Waste(s)	Benefits to Water	Benefits to Other Media	Other Incentives	Data Source
Source Reduction and Recycle/Reuse: Use of activated carbon to decolorize wastewater and subsequent reuse for raw feed water	Dyeing	Dye	By removing dye from wastewater, it may be more readily reused as raw water feed. This results in a reduction of the quantity of wastewater produced by 80 percent.		Reduces costs related to water usage and purchase of dyes	T6
Source Reduction and Recycle/Reuse: Wastewater reuse and water conservation	Sizing, dyeing, and printing	Biological oxygen demand	Conserves water, reduces amount of wastewater generated, and reduces BOD in dye department wastewater by 20 percent			T7
Source Reduction: Chemical substitution for sodium sulfide	Black color dyeing	Sodium sulfide	Reduces sodium sulfide in effluent to meet effluent limit		Eliminates foul smell of sulfide	T8
Source Reduction and Recycle/Reuse: Installation of automatic water stops	Dyeing	Water	Conserves water use by 25 to 79 percent		Reduces cost of effluent treatment	T9
Source Reduction and Recycle/Reuse: Introduction of a fourth washing stage using a pressure washer	Washing and bleaching synthetic fibers	Liquid impurities in washwaters that required a large amount of chlorine consumption in the bleaching process	Reduces COD by 22 percent, chlorinated organic compounds by 14 percent, and color by 40 percent in effluent discharged		Reduces steam energy requirement as well as amount of chlorine and makeup alkali needed	T11
Recycle/Reuse: Reuse of both non-contact cooling water and contact production water	Processing dye batches of yarn	Typical process pollutants and water	By storing the heated non-contact cooling water in hot storage tanks, this water became available for reuse in coloring. This reduces wastewater generated by 60 percent.		Resulted in water savings of approximately $13,000 per month	T11
Recycle/Reuse: Reuse of heated water	Dye liquor preparation	Water	Reuses heated wastewater that was at one time lost to effluent		By using waste heat, steam and energy requirements were reduced, and fuel use was reduced by approximately 440 gallons per day	T11

EXHIBIT 3-4. SUMMARY OF BMPs UTILIZED IN THE TEXTILES MANUFACTURING INDUSTRY (Continued)

BMP	Targeted Process(es)	Targeted Waste(s)	Benefits to Water	Benefits to Other Media	Other Incentives	Data Source
Recycle/Reuse: Use of automated control to reuse water	Water cooling	Water	Pressure valves allow water to be reused from onsite tanks when available. This bypasses the city water line and reduces the need for fresh water.		Reduces water costs, and automatic controls of tank levels reduce labor costs	T11
Source Reduction: Use of computer programs to accurately control dye absorption	Dyeing	Dyes and water	Process modifications reduce excess dyes that significantly eliminate the need for rinsing after dyeing		Efficient use of raw materials reduces expenditures for dyes and water	T11
Recycle/Reuse: Use of multistage countercurrent wash system to produce a more concentrated effluent stream for recycling	Washing	Water, aqueous wastes	By reducing the amount of water in the wastestream, the washwater wastestream may be easily concentrated and both pollutants and water recycled			T12
Recycle/Reuse: Redesign of washers	Washing	Water	By designing washers to use only the amount of water necessary for a particular step or operation, water use decreases. Up to 85 percent less water may be used with new washers.			T12
Source Reduction: Use of an ultraviolet light disinfection unit instead of biocides to control microbial growth in cooling water	Cooling	Chemical biocides	Use of ultraviolet disinfection avoids the need to introduce chemical biocides which would enter the wastewater			T13
Source Reduction: Pre-wetting of fabric to reduce the demand for urea	Dyeing and washing	Ammonia nitrogen	Since less urea was needed, the formation and discharge of ammonia nitrogen was lessened		Allowed the facility to meet state effluent regulations for ammonia nitrogen	T13
Source Reduction: Replacement of alkyl phenol ethoxylates with linear alcohol ethoxylate compounds	Hosiery mill manufacturing	Alkyl phenol ethoxylates	Eliminated the more toxic alkyl phenol ethoxylates from the wastewater discharge		Saves approximately $2,000-$5,000 per month in chemical costs	T13

EXHIBIT 3-4. SUMMARY OF BMPs UTILIZED IN THE TEXTILES MANUFACTURING INDUSTRY (Continued)

BMP	Targeted Process(es)	Targeted Waste(s)	Benefits to Water	Benefits to Other Media	Other Incentives	Data Source
Source Reduction: Use of automated chemical handling system	Chemical and dye dispensing	Textile dyes and chemicals	Allows for the introduction of accurate amounts of chemicals or dye to machines, thereby eliminating the risk of human error and reducing the amount of chemicals and dye in the wastewater discharge		By using the automated chemical handling system, fewer error and batch dumps, and fewer safety problems from handling occur. Better repeatability for lab to dye house correlation also resulted.	T13
Source Reduction: Use of temperature controls to reduce need for retarders, levelers, and defoamers	Line operation	Retarders, levelers, defoamers	Lessens amount of retarders, levelers, and defoamers used, thus reducing the amount of toxics in the wastewater discharge		Conservation of chemicals results in savings to the facility	T13
Source Reduction: Use of hyper-filtration to recover caustic from spent solutions	Wool scouring	Caustics	Allows for the recovery and reuse of caustics, thereby minimizing such discharges			T14
Recycle/Reuse: Reuse of final rinse as makeup for the next bath	Finishing	Heavy metals	Reduces water use and recovers heavy metals			T15
Source Reduction: Use of automatic shut off valves, flow indicators, and flow meters	All processes	Typical process pollutants	Helps to quickly identify spills and respond to problems which cause water loss			T15
Source Reduction: Use of dyes with less toxic pollutants and less water content	Dyeing	Toxics in dyes and water	Improves dyeing by reducing water content in dyes and by substituting less hazardous dyes and dye carriers. This reduces the discharge of pollutants and reduces water discharge			T15
Source Reduction: Use of jet dyeing and low liquor dyeing	Dyeing	Dyes and water	Increases the efficiency of dyeing and reduces water consumption			T15

EXHIBIT 3-5. SUMMARY OF BMPs UTILIZED IN THE PULP AND PAPER MANUFACTURING INDUSTRY

BMP	Targeted Process(es)	Targeted Waste(s)	Benefits to Water	Benefits to Other Media	Other Incentives	Data Source
***Recycle/Reuse*:** Installation of a closed system	All processes	All pollutants in the wastewater	A closed system blends in fresh water only as makeup water or as dilution for some component that might otherwise restrict recycling. This results in the wastewater discharge being eliminated.	The amount of solids generated is reduced from 6 percent to 1 percent of waste production		P1
***Source Reduction*:** Use of chlorine, caustic soda, and chlorine dioxide bleaching preceded by oxygen bleaching	Bleaching	Dioxin and dioxin precursors	Reduces the quantities of reagents and water used in the conventional bleaching process, resulting in a decrease in the amount of water needed by 50 percent	Other feedstock reductions include an 18 percent reduction in chlorine, a 29 percent reduction in soda, and a 33 percent reduction in chlorine dioxide. Also reduces the coloration of wastes by 50 percent.	Savings of 9.1 FF per ton in raw materials and water consumption	P1
***Source Reduction*:** Installation of dry bark-stripping technology	Bark stripping	Solids, biochemical oxygen demand	Virtually eliminates wastewater — the conventional process uses 9 cubic meters per ton of pulp produced while the dry barking process uses 0.25 cubic meters per ton of pulp produced	Solid wastes that are generated can be used as a fuel source without further treatment	The capital investment of the dry barking process is 25 percent less than that of the conventional technology (wet bark-stripping); operation and maintenance costs are 33 percent less. Reduction in waste requiring disposal corresponds to a reduction in disposal costs. The conventional process uses 117 MJ of electricity per ton of pulp produced while the clean technology uses 74.6 MJ per ton of pulp produced.	P3
***Recycle/Reuse*:** Installation of spill pits to catch process water and route for reuse	Papermaking	All pollutants	Allows for reuse of process water			P2

EXHIBIT 3-5. SUMMARY OF BMPs UTILIZED IN THE PULP AND PAPER MANUFACTURING INDUSTRY (Continued)

BMP	Targeted Process(es)	Targeted Waste(s)	Benefits to Water	Benefits to Other Media	Other Incentives	Data Source
Source Reduction: Improvement in the washing of brownstock	Washing	Dioxin, dioxin precursors, and biochemical oxygen demand	By improving the brownstock washing process, the precursors that form dioxin are minimized. This helps to reduce the toxicity of wastewater and allows for the achievement of regulatory compliance.			P2
Source Reduction: Use of additives, such as defoamers	Bleaching and papermaking	Dioxin	Additives such as defoamers minimize the potential to form dioxin when the additives are exposed to chlorine. This helps to reduce the toxicity of wastewater and allows for the achievement of regulatory compliance.			P2
Recycle/Reuse: Employment of used heated water for a multiple pass system instead of a single pass system	Papermaking		Reduces water usage and the amount of water that must be treated		Reduces fuel consumption and associated costs, as well as water usage and treatment costs	P2
Source Reduction: Substitution of chlorine dioxide for chlorine in the first stage of bleaching	Bleaching	Dioxin and dioxin precursors	Use of chlorine minimizes the formation and subsequent discharge of chlorinated organics			P3
Source Reduction: Substitution of uncontaminated defoamers and pitch dispersants which contain no dibenzofurans and dioxin precursors	Pulping	Dioxin, dioxin precursors, and dibenzofurans	Use of defoamer and pitch dispersents minimizes the formation and subsequent discharge of chlorinated organics			P4
Source Reduction: Extended delignification using the kraft cooking or polysulfide cooking processes	Pulping	Dioxin and dioxin precursors	Reducing the need for leaching chemicals helps minimize the amount of chlorinated organics formed and discharged		Realized a return on investment in less than 1.5 years	P4

EXHIBIT 3-5. SUMMARY OF BMPs UTILIZED IN THE PULP AND PAPER MANUFACTURING INDUSTRY (Continued)

BMP	Targeted Process(es)	Targeted Waste(s)	Benefits to Water	Benefits to Other Media	Other Incentives	Data Source
***Source Reduction*:** Improvement in brownstock washing using pressure washers	Washing	Dioxin and dioxin precursors	Improved pulp washing reduces the amount of chlorine required for bleaching, thereby minimizing the formation and discharge of chlorinated organics		Creates substantial energy savings; requires minimal costs to optimize existing systems; realizes cost savings due to decreased use of bleach	P4
Recycle/Reuse: Use of spill containment techniques	Spills	Unbleached pulp fiber, waste liquor, cooking liquor	Overflow compartments, collection sumps, collection pumps and storage tanks prevent wastes from reaching water systems and facilitate the return of spills to production processes		Recovered materials reduce costs of raw materials and reduce wastewater treatment costs	P5
Source Reduction: Enclosing brownstock screen room	Vacuum washing	Color and residual solids	Contains pollutants including color and reduces effluent volume			P5
Recycle/Reuse: Optimization of condensate treatment system, steam stripping, and wastestream segregation	Wood cooking	Contaminated evaporator condensates, digester vent and blow condensates	Allows for the recovery and reuse of evaporator condensates, digester vent, and blow condensates. This eliminates the necessity to discharge condensates.		Reuse and recovery reduces raw material costs	P5
Recycle/Reuse: Use of new equipment such as piping, tanks, pumps and cooling towers	Paper making	Typical process pollutants	Segregation and collection of clean cooling and service water can minimize water requirements and facilitate reuse. Contaminated effluents in wastewater are minimized.		Reduces raw water costs	P5
Source Reduction: Use of process modification	Bleaching systems	Color and bleaching effluent pollutants	Countercurrent leached pulp washing reduces chemical requirements for color removal. This process reduces the load of bleaching wastes in wastewater.		Reduces raw material costs for lime and alum	P5
Recycle/Reuse: Installation of high pressure oscillating showers to replace stationary showers	Kraft pulp, board manufacture and pulp drying	Typical process pollutants	The new equipment minimizes water requirement and improves cleaning. Separators segregate white water and seal water to facilitate recovery and reuse.		Returns recovered pulp to the process	P5

EXHIBIT 3-5. SUMMARY OF BMPs UTILIZED IN THE PULP AND PAPER MANUFACTURING INDUSTRY (Continued)

BMP	Targeted Process(es)	Targeted Waste(s)	Benefits to Water	Benefits to Other Media	Other Incentives	Data Source
Recycle/Reuse: Reuse of rich white water for stock preparation and dilution prior to treatment	Paper making	Water	Eliminates the need for fresh water addition during pulping and facilitates pulp recovery from wastewater			P5
Recycle/Reuse: Use of service water clean save-all filtrate in the white water system	Paper making	Water	Use of fresh water to the white water system can be replaced by using service water or save-all filtrate			P5
Recycle/Reuse: Use of vacuum boxes and rolls equipped to facilitate waste segregation	Paper making	Clean seal water	Allows white water to be sent to save-all and clean water seal water can be recovered and collected for reuse as shower water			P5
Source Reduction: Installation of a dry barking system	Bark removal	Solids, biological oxygen demand	Eliminates the need for pressing and bark thawing. These modifications eliminate bark removal process discharges.			P5
Process Modification: Separation of chlorine bleaching activities from refining and cleaning	Unbleached krafting	Screening extracts and cleaner rejects	Allows extract from screening and cleaner rejects from bleachable grade pulp mills to be returned to the process rather than discharged to the sewer system			P5
Source Reduction: Use of screen filtering and rewashing techniques	Cooking liquor manufacturing	Carbonate, caustic material, and grits	Disposable solids can be washed over a small filter to recover and reuse filtrates and reusable chemicals. Eliminates a source of biochemical oxygen demand in the wastewater.			P5
Source Reduction: Use of diffusion washing	Paper making	Water	Reduces wash water requirement and bleaching effluent flow by entering pulp into the washer at higher densities		Provides fuel and chemical savings	P5
Source Reduction: Use of gas phase bleaching	Bleaching	Water	Use of gaseous bleaching reagents on pulps at high densities reduces effluent flow		Reduces total reaction time from 10 to 2 hours. This process also reduces chemical and heat costs.	P5

EXHIBIT 3-5. SUMMARY OF BMPs UTILIZED IN THE PULP AND PAPER MANUFACTURING INDUSTRY (Continued)

BMP	Targeted Process(es)	Targeted Waste(s)	Benefits to Water	Benefits to Other Media	Other Incentives	Data Source
Source Reduction: Use of rapid displacement heating to reduce the amount of chlorine needed in pre-bleaching	Cooking	Chlorine, chlorinated organics	Allows for the extension of the kraft cooking process and produces low lignin pulp. This results in the need for less chlorine and reduces the chlorine available for chlorinated organic formation which would enter the wastewater discharge.		Results in a 10 percent increase in pulp strength, a 10 to 15 percent reduction in pulping cycle time, a 65 percent reduction in steam demand, and a 2 to 3 gigajoules/ADMT energy savings	P6
Source Reduction: Use of oxygen delignification to reduce amount of chlorine needed in bleaching	Bleaching	Chlorine, chlorinated organics, biochemical oxygen demand, chemical oxygen demand, and color	Reduces the amount of chlorine conventionally required by 50 percent. Decreases the amount of chlorinated organics, biochemical oxygen demand, chemical oxygen demand, and color contained in the effluent discharge.		Oxygen delignification pulps are reported to be equal or superior to conventionally bleached pulps with respect to tear strength, brightness stability, pitch removal, and cleanliness	P6
Source Reduction: Replacement of chlorine gas with chlorine dioxide to reduce formation of chlorinated organics, dioxins, and furans	First bleaching stage	Chlorinated organics, dioxins, and furans	Reduces the amount of chlorinated organics, dioxins, and furans formed, thus reducing the levels of these contaminants in the wastewater discharge		Chlorine dioxide substitution resulted in lower bleach plant chemical consumption, and lower bleach plant costs	P6

EXHIBIT 3-6. SUMMARY OF BMPs UTILIZED IN THE PESTICIDES MANUFACTURING INDUSTRY

BMP	Targeted Process(es)	Targeted Waste(s)	Benefits of Water	Benefits to Other Media	Other Incentives	Data Sources
Source Reduction and Recycle/Reuse: Use of dedicated baghouses on each formulation mill and enclosing unit operations and process equipment	Formulation	Commercial pesticides	Reduces wastewater use in that drums are no longer washed. Allows for capturing of baghouse dust which may be recycled, reusing raw diluent containers, and recycling of empty drums.	Solid waste volume is reduced by nearly 80 percent		S1
Source Reduction and Recycle/Reuse: Use of steam-cleaning mixing tanks instead of batch-boil techniques, installation of low-volume water spray nozzles on rinsing and cleaning equipment, and installation of recycling equipment cleaning water for process feed water	Equipment cleaning	Commercial pesticides	Minimizes and recycles equipment rinsewater as a formulation diluent to thereby reduce the amount of pesticides entering the wastewater as well as the amount of wastewater generated			S1
Source Reduction and Recycle/Reuse: Elimination of pesticide rinse through reuse of spent rinse after dilution	Rinsing	Commercial pesticides	Eliminates the discharge of rinsewater contaminated with pesticides			S2
Source Reduction and Recycle/Reuse: Use of wipers and squeegees to remove residual on mix tank walls after tanks are drained	Equipment cleanup, rinsing	Commercial pesticides	Removing tank residues with wipers and squeegees reduces the amount of rinsewater needed for cleanup and reduces the amount of pollutants that come into contact with rinsewaters. This decreases the quantity of wastewater discharged and lessens the pollutant load.	Waste rags are no longer needed and do not pose a disposal problem	Reduces water usage, wastewater treatment, and disposal resulting in lower costs	S3
Source Reduction and Recycle/Reuse: Installation of high pressure sprays for equipment cleanup using nozzles on all hoses	Equipment cleanup	Commercial pesticides	High pressure sprays use less rinsewater than conventional sprays. Water consumption is cut by 80 to 90 percent.		Reduces water usage, resulting in lower costs	S3

EXHIBIT 3-6. SUMMARY OF BMPs UTILIZED IN THE PESTICIDES MANUFACTURING INDUSTRY (Continued)

BMP	Targeted Process(es)	Targeted Waste(s)	Benefits of Water	Benefits to Other Media	Other Incentives	Data Sources
Source Reduction and Recycle/Reuse: For spill cleanup, the use of dedicated vacuums for dry spills and dedicated mops and squeegees for liquid spills; use of recycled water where water is needed for cleanup	Equipment cleaning	Commercial pesticides	Using dedicated cleanup equipment allows for easier recovery of materials. These methods of cleanup also reduce wastewater volume associated with cleanup, and lessen contamination of wastewater.	Dry cleanup creates less solid waste disposal		S3
Source Reduction: Extension of production runs of the same product for as long as possible	Chemical blending	Commercial pesticides	Longer manufacturing sequences of the same product or family of products reduces the amount of cleanout required and thus minimizes wastewater discharge quantities		Decreases costs associated with water usage and wastewater treatment	S3
Recycle/Reuse: Storage and reuse of rinsewater; use of reused rinsewater as the initial rinse when more than one rinse is required	Equipment cleanup	Commercial pesticides	Reusing rinsewater reduces the quantity of wastewater discharged		Decreases costs associated with water usage and wastewater treatment	S3
Source Reduction and Recycle/Reuse: Segregation of hazardous from nonhazardous waste, and of spent chlorinated from nonchlorinated solvents for offsite recovery; return of unused agricultural research chemicals for reuse or reformulation	Laboratory research	Commercial pesticides	Reduces pesticides in wastewater by preparing chlorinated and nonchlorinated solvents for offsite recovery and returning unused agricultural research chemicals for reuse or reformulation	Hazardous waste generation is reduced by 70 percent		S4
Recycle/Reuse: Reuse of rinsewater as make-up water, diluent, or carrier during the next formulation of the same product	Equipment cleaning	Commercial pesticides and water	Reduces the amount of water used and pesticides discharged			S5
Source Reduction: Rinsing of drums using high pressure spray system	Equipment cleaning	Water	Reduces amount of pesticide contaminated rinsewater			S5

EXHIBIT 3-7. SUMMARY OF BMPs UTILIZED IN THE PHARMACEUTICAL MANUFACTURING INDUSTRY

BMP	Targeted Process(es)	Targeted Waste(s)	Benefits to Water	Benefits to Other Media	Other Incentives	Data Source
***Recycle/Reuse*:** Use of waste filter cakes as a soil additive	Filter cake management	Waste filter cakes	Use of filter cakes as a soil additive eliminates the disposal into the sewer		Will reduce the estimated disposal costs of filter cakes at full-scale production of approximately $250,000 per year	H1
***Recycle/Reuse*:** Installation of on-site recovery equipment of spent solvent solutions operations	Equipment cleaning and chemical reactions in production	Spent solvent solutions of amyl acetate and acetone	Recovery operations result in the reuse of more than 99 percent of solvents processed, which minimizes disposal needs		Based on a cost of $1.78 per gallon of raw solvent, saves $3,520 to $5,290 in raw material costs per harvest	H1
***Source Reduction*:** Installation of guides beneath rotary vacuum filters that direct filter cake into the center of the conveyor belt	Filter cake management	Waste filter cakes	New equipment will reduce disposal loads		Since the filter cakes may have some value as a soil additive, efficient collection of filter cake may be financially beneficial	H1
***Recycle/Reuse*:** Reclamation of alcohol-based wastes and mixing them with fuel oil, which can be used as fuel for a boiler	Waste disposal	Alcohol-based wastes	Eliminated disposal of alcohol-based wastes		Company saved 1,000 gallons of fuel oil per year ($2,800). Eliminated offsite disposal costs.	H2
***Recycle/Reuse*:** Installation of a separation process to recover and reuse acetone from wastewater	Antibiotics manufacturing	Acetone	Recovered and reused 70 percent acetone by weight from wastewater		New process saves $70,000 annually in treatment costs	H3
***Source Reduction and Recycle/Reuse*:** Use of vacuum systems for routine clean-up of dry chemical spills rather than water	Cleaning	Typical process pollutants	Vacuum collection devices reduce the amount of clean-up water needed and lessens the pollutant load and the volume of water entering the wastestream			H4
***Source Reduction and Recycle/Reuse*:** Use of squeegees and vacuum collection devices rather than water to clean up spills	Cleaning	Typical process pollutants	Vacuum collection devices and squeegees when cleaning up reduce the amount of clean-up water needed. This lessens the pollutant load and the volume of water entering the wastestream.			H4

EXHIBIT 3-7. SUMMARY OF BMPs UTILIZED IN THE PHARMACEUTICAL MANUFACTURING INDUSTRY (Continued)

BMP	Targeted Process(es)	Targeted Waste(s)	Benefits to Water	Benefits to Other Media	Other Incentives	Data Source
***Recycle/Reuse*:** Segregation of arsenic-laden wastestreams, concentration of these wastestreams, and recovery of arsenic	Manufacturing	Arsenic	Segregation and concentration of arsenic-laden wastestreams enables the arsenic to be reused rather than discharged		Reduces expenditures for wastewater treatment and raw materials	H4
***Recycle/Reuse*:** Reuse of once through non-contact cooling water as waste combustion scrubber water, reuse of deionized rinsewater as cooling tower makeup, and collection of effluent cooling water in a pond and use as fire protection	Manufacturing	Typical process chemicals	Reuse of cooling waters and rinsewaters reduces the volume of wastewater discharge and reduces water usage		Reduces costs associated with water usage	H4
***Recycle/Reuse*:** Implementation of a recovery project	Phosphate ester reactions	Phenols	Recovers phenol from wastewater for reuse in succeeding batches		Reduces phenol purchasing costs and sludge disposal costs	H5

EXHIBIT 3-8. SUMMARY OF BMPs UTILIZED IN THE PRIMARY METALS MANUFACTURING INDUSTRY

BMP	Targeted Process(es)	Targeted Waste(s)	Benefits to Water	Benefits to Other Media	Other Incentives	Data Source
***Source Reduction and Recycle/Reuse*:** Installation of closed-loop system at an iron casting foundry	All processes	Typical process pollutants	Conserves water. Only 25 percent of the total process water has to be treated. Also reduces capital costs for the pretreatment facility by 50 to 75 percent.	Conserves wastewater treatment chemicals, reduces pumping costs, and decreases health risks due to less exposure to treatment chemicals	Added $18,744 per year in operation and maintenance costs. Saves $13,000 per year in feedstock reduction, $21,000 per year due to waste reduction, and $.25 per ton produced including water and energy costs.	M1
***Recycle/Reuse*:** Use of chemical precipitation in a carbon steel wire manufacturing plant	Carbon steel wire manufacturing	Iron, lead, and zinc	Comprehensive wastewater system produces finished water suitable for recycle and reuse in the manufacturing process, thereby conserving water		Reduced water and sewer rates by $5,400 per month	M2
***Recycle/Reuse*:** Use of rinsewater	Rinsewater	Copper, nickel	Allows for 90 percent recovery and reuse of the rinsewater		Copper and nickel, both subject to local discharge limits (2.07 milligrams per liter for copper and 2.38 milligrams per liter for nickel) were increasingly costly to discharge. Using the new system, the remaining discharge is within local limits.	M3
***Recycle/Reuse*:** Use of spent process pickle liquor	Process water	Ferrous chloride and hydrogen chloride	Reuses 13,500 gallons per day of reclaimed water in steel operations	Dilutes 20 to 25 tons of ferrous chloride to 30 percent solution and ships to market. Returns 3,550 gallons per day of hydrogen chloride to the steel operation.	Minimizes corporate liability through less waste disposal and provides continual revenue (not quantified) from sale of ferrous chloride to local industries	M4

EXHIBIT 3-8. SUMMARY OF BMPs UTILIZED IN THE PRIMARY METALS MANUFACTURING INDUSTRY (Continued)

BMP	Targeted Process(es)	Targeted Waste(s)	Benefits to Water	Benefits to Other Media	Other Incentives	Data Source
***Source Reduction*:** Use of alcohol instead of acid in copper pickling operation	Pickling operations, rinsing, acid dumps	Typical process pollutants	Eliminates rinsewater discharge. Former technology required 750 liters of rinsewater and 0.75 kilograms of effluent-neutralizing products.		Process can be applied to continuous casting and drawing units; only feasible when surface oxidation is light and when the copper is at a temperature high enough for chemical reaction to occur Capital costs are 300,000 francs; operation and maintenance costs are 8.55 francs per ton of pickled wire	M5
***Source Reduction and Recycle/Reuse*:** Use of cyanide-free process baths, water reuse, filtration and monitoring, longer drip times, spray rinsing, closed cooling, and wastewater treatment	Plating and rinsing	Cyanide, acid, chrome, iron, and zinc	Metals in effluent reduced from 945 kilograms per year to 37 kilograms per year. Water use reduced from 330,000 cubic meters per year to 20,000 cubic meters per year (including 3,500 cubic meters per year of noncontaminated cooling water used to prevent foaming in the end control pit).		Company realized tax savings as a result of reduced heavy metal pollution	M6
***Source Reduction*:** Use of drip plates, changes in rinsing procedures, and installation of a sedimentation tank	Rinsing	Chrome, nickel, and zinc	Estimated dragout reduction of 95 percent. Generated 50 percent less wastewater (reduced from 16,000 to 8,000 cubic meters per year). Metals reduced from 33-200 to 16 kilograms per year (report states 33 kilograms per year but individuals at the plant estimated the original figure to be 100-200 kilograms per year).		Investment costs were Dfl 70,000; capital costs were Dfl 14,000 per year. Annual operations and maintenance costs (in Dfl): labor 10,000; chemicals & energy 2,000; sludge removal 1,500; in process measures 10,000.	M6
***Recycle/Reuse*:** Installation of an electrolytic cell to recover zinc in low concentration iron-containing rinsewaters	Pickling operations, plating, rinsewaters	Zinc and iron	Reduced total zinc dragout by 86 percent (from 980 to 140 grams per hour) with an extra energy requirement of 6.4 kilowatts	Avoided landfilling costs, saving $19,440 per year	Investment costs were $180,000 for equipment and installation with an expected life of 10 years; operation and maintenance costs are $23,040 per year	M7

EXHIBIT 3-8. SUMMARY OF BMPs UTILIZED IN THE PRIMARY METALS MANUFACTURING INDUSTRY (Continued)

BMP	Targeted Process(es)	Targeted Waste(s)	Benefits to Water	Benefits to Other Media	Other Incentives	Data Source
***Source Reduction*:** Substitution of degreasing and painting operations with iron phosphating and powder coating processes	Degreasing and painting operations	Trichloroethane	Use of powder-coating or iron phosphate final product finish avoids the need for degreasing with trichloroethylene. Also, new wastewater treatment plant reduces wastewater discharges.	Reduced chemical consumption per feedstocks; realize energy savings; increased productivity, improved working environment, and produced a better quality product	Investment cost was 700,000 kroner; operation and maintenance costs were 45,500 kroner	M8
***Source Reduction and Recycle/Reuse*:** Use of acid purification unit and an ion exchanger on concentrated high temperature pickling liquor to reduce generation of iron	Pickling operations	Iron		Reduced the iron content of the acid solution from 7.7 to 2-3 percent. Also reduced feedstock use (sulfuric acid by 89 percent and lime by 89 percent).	Capital costs were $96,500; payback period, 2.33 years. Operation and maintenance costs were $2,500 per year. Estimated $8,000 savings on sludge hauling. Annual savings on chemicals was $43,937.	M9
Source Reduction: Use of dry air pollution devices	Air emission control	Gaseous pollutants commonly found in aluminum forming	Dry air emission devices generate no wastewater stream		Dry air pollution control devices may be more cost effective than wet devices	M10
Source Reduction: Maintenance of equipment such as pumps, piping and presses	All operations	Hydraulic oils and pollutants found in process chemicals, filter cleaning waters, and washdown water	Proper equipment maintenance reduces wastewater loads by minimizing leaks, spills and other problems related to equipment failure			M10
Recycle/Reuse: Use of alternative rinsing techniques such as countercurrent cascade rinsing and spray rinsing	Rinsing	Pollutants commonly found in aluminum forming rinsewaters	Reduces the amount of water necessary to achieve required cleanliness of the work piece. Also improves wastewater treatment efficiency.	Efficient rinsing reduces the need for chemical treatment	Efficient use of water decreases water costs	M10

EXHIBIT 3-8. SUMMARY OF BMPs UTILIZED IN THE PRIMARY METALS MANUFACTURING INDUSTRY (Continued)

BMP	Targeted Process(es)	Targeted Waste(s)	Benefits to Water	Benefits to Other Media	Other Incentives	Data Source
Recycle/Reuse: Regeneration using temperature changes, chemical addition, and ulrafiltration	Metals forming	Various chemical etching and cleaning baths	Removes contaminants to allow for recovery and reuse of bath chemicals, thereby reducing the volume of bath water discharge		Increases efficiency of cleaning and etching operations and reduces costs for labor and chemical purchases	M10
Recycle/Reuse: Reclamation of oily wastes available as a result of chemical emulsion breaking	Rolling and drawing	Emulsions	Breaks down stable oil in water emulsions to allow for the reclamation of oily wastes from wastewater			M11
Recycle/Reuse: Use of heat to destabilize oil droplets in spent emulsions to separate oil, distilled water, sludge and other floating materials	Aluminum forming	Emulsions	Thermal emulsion breaking uses heat to break down spent emulsions. One product of this process is distilled water that can be reused. At least 99 percent oil from the wastewater is removed.	Separates floating oil from sludge solids thereby reducing amount of sludge to be disposed	The automation in the operation results in reduced labor costs	M11
Recycle/Reuse: Use of carbon adsorption to remove dissolved organics from wastewater	Wastewater treatment	Organics and oils	Allows for the practical recovery of adsorbed materials and removes 65 percent of toxic organics from wastewater		Applicable to a wide range of organics. The system is compact, and tolerates a wide range of rates and concentrations.	M11
Recycle/Reuse: Use of coalescence to remove finely dispersed oil droplets to e reused	Wastewater treatment	Oily wastes	Removes oils from wastewater that are too finely dispersed for gravity separation, thereby allowing reuse		Reduces residence times for achieving separation of oil from wastewater	M11
Recycle/Reuse: Use of evaporation for concentrating and recovering process solutions	Wastewater treatment	Process solutions and phosphate metal cleaning solution pollutants	Recovers process and distilled water for rinsing			M11

EXHIBIT 3-8. SUMMARY OF BMPs UTILIZED IN THE PRIMARY METALS MANUFACTURING INDUSTRY (Continued)

BMP	Targeted Process(es)	Targeted Waste(s)	Benefits to Water	Benefits to Other Media	Other Incentives	Data Source
Recycle/Reuse: Use of ion exchange system	Wastewater treatment	Aluminum, arsenic, cadmium, chromium, copper, cyanide, gold, iron, lead, manganese, nickel, selenium, silver, tin, and zinc	Efficiently recovers metals from wastewater		Ion exchange is a flexible technology that is applicable to many situations	M11
Recycle/Reuse: Use of reverse osmosis to separate wastewater stream into concentrated and dilute streams and return them to the process to form a closed-loop system	Rinsing	All pollutants in rinsewater	Eliminates discharges of rinsewater		Reverse osmosis concentrates dilute solutions for recovery of salts and chemicals with low power requirements. Reduced cost of fresh water input.	M11
Recycle/Reuse: Recirculation of water from one process into a different production process	Extrusion die cleaning	Wastewater	Reduces water consumption which allows wastewater treatment to be designed for smaller flows			M11
Recycle/Reuse: Regeneration and reuse of chemical baths by precipitating out waste metals.	Cleaning, etching, coating, and anodizing	Chemicals associated with mentioned processes	Results in zero discharge of bath water by reusing all spent baths that have been sent through ultrafiltration membranes		Cleaning and etching operations are made more efficient because baths are kept at a constant strength. Reduces bath maintenance labor costs. Reduces chemical costs by recovering chemicals and increasing bath life.	M11
Recycle/Reuse: Use of a pressure filter to recover wastewater	Continuous casting	Oil, solids, metals	Can return 96 percent of the filtered effluent to the process			M12
Recycle/Reuse: Use of a vacuum filter to recover wastewater	Steel making	Suspended solids	Results in opportunities to recycle of a major portion of the effluent to the process			M12

EXHIBIT 3-8. SUMMARY OF BMPs UTILIZED IN THE PRIMARY METALS MANUFACTURING INDUSTRY (Continued)

BMP	Targeted Process(es)	Targeted Waste(s)	Benefits to Water	Benefits to Other Media	Other Incentives	Data Source
Recycle/Reuse: Use of thickeners, cooling towers and dewatering processes to recover wastewater	Iron making	Blast furnace wastewater, dewatered solids	Allows for recycle of 90 percent of blast furnace wastewater	Allows for recovery of dewatered solids for use in sintering operations		M12
Recycle/Reuse: Use of a reverse osmosis system	Anodizing process	Caustic and acid solutions	Allows for the recovery and reuse of wastewater		Improves raw material recovery and realizes savings for water and raw material purchases	M13
Recycle/Reuse: Use of an electrodialysis unit	Anodizing process	Aluminum hydroxide	Removal of aluminum hydroxide from wastewater allows for the wastewater to be reused		Improves raw material recovery and realizes savings for water and raw material purchases	M13
Recycle/Reuse: Use of air agitation units and reduction of flow rates of water through the rinse tanks	Rinsing	Wastewater	Increases rinsing effectiveness and reduces water needs		Reduces water purchases	M13
Recycle/Reuse: Installation of a closed-loop, chilled-water system using recirculated water to cool ultraviolet oven and etch tanks	Cooling and rinsing	Water	Reduced wastewater discharged by 60 percent		Resulted in a payback period of 1.9 years	M14
Recycle/Reuse: Filtration of bath water containing caustic soda beads and etch resist ink particles	Rinsing	Etch resist ink particles	Allows for the removal of etch resist ink particles			M14
Recycle/Reuse: Reuse of water from waste treatment system as high pressure spray rinse	Rinsing	Water	By reusing water from the treatment system as high pressure spray rinse, water usage deceases			M14
Recycle/Reuse: Removal of metal contaminants from rinsewater	Rinsing	Water	Makes rinsewater available for reuse			M14

EXHIBIT 3-8. SUMMARY OF BMPs UTILIZED IN THE PRIMARY METALS MANUFACTURING INDUSTRY (Continued)

BMP	Targeted Process(es)	Targeted Waste(s)	Benefits to Water	Benefits to Other Media	Other Incentives	Data Source
Recycle/Reuse: Reuse of cascade rinsewater overflow to replace tap water make up	Rinsing	Water	Saves 650,000 gallons of tap water		Savings result since 650,000 gallons of water does not have to be purchased and discharged to the sewer. Provides a payback period of 0.4 years.	M14
Source Reduction: Installation of flow reducers and flow meters on the metal cleaning line	Cleaning	Water	Decreases water use by 124,800 gallons per year		Provides a payback period of 0.6 years	M14
Recycle/Reuse: Installation of reverse osmosis solution recovery system, electrodialysis unit, in-tank air agitation units, and lowering of the flow rate of water through the rinse tanks	Anodizing process line	Wastewater	Allows for the recovery and recycle of the raw materials lost in the rinsing operations. Reduced water purchases by 85 percent.		Provides a payback period of 4 years	M14
Source Reduction: Substitution of a water soluble synthetic with 1,1,1-trichloroethane	Degreasing	Trichloroethane	Eliminates the discharge of trichloroethane		Saves approximately $12,000 per year and reduces waste disposal costs by over $3,000 annually. Also, plant personnel now experience improved health and safety conditions at the plant.	M15

EXHIBIT 3-9. SUMMARY OF BMPs UTILIZED IN THE PETROLEUM REFINING INDUSTRY

BMP	Targeted Process(es)	Targeted Waste(s)	Benefits to Water	Benefits to Other Media	Other Incentives	Data Source
Source Reduction: Use of air cooling system instead of water cooling systems	Petroleum refining cooling	Cooling water pollutants	Increased use of air cooling systems reduces the quantity of cooling tower blowdown discharges that require treatment			R1
Source Reduction: Elimination of cooling water from general purpose pumps	Petroleum refining pumping	Cooling water pollutants	Reduces quantity of wastewater		Elimination of water can increase machinery reliability, reduce expenses for piping and water treatment, and save operating costs	R1
Recycle/Reuse: Use of treated wastewater as makeup to the cooling tower and fire water systems	Cooling	Phenols	Cooling tower acts as a biological treatment unit that removes 99 percent of phenols from the water. The refinery reuses 4.5 million gallons of water per day in the cooling tower.			R1
Recycle/Reuse: Use of skimming	Wastewater treatment	Oil	Removes oil from the wastewater for recycle		Recovered oil can be treated and made ready to sell	R2
Source Reduction: Installation of new tanks to segregate oil, water and solids	Wastewater treatment	Oil, water, and solids	Provides mechanism for separation of oil, water and solids. Segregated wastes facilitate recovery and reuse.	Facilitates treatment of sludge		R2
Recycle/Reuse: Washing of entrapped jet fuel from spent treater clay to recover jet fuel	Filtration though clay towers	Jet fuel	Minimizes jet fuels contained in discharges		Greatly decreased disposal costs. Allows for the recovery and recycle of approximately 2,000 barrels per year of jet fuel.	R3

EXHIBIT 3-11. SUMMARY OF BMPs UTILIZED IN THE INORGANIC CHEMICAL MANUFACTURING INDUSTRY

BMP	Targeted Process(es)	Targeted Waste(s)	Benefits to Water	Benefits to Other Media	Other Incentives	Data Sources
***Recycle/Reuse*:** Routing of drip acid for reuse in the reactor	Hydrofluoric acid production	Fluorosulfonic acid	Recycling drip acid to the reactor avoids having to send it to the wastewater treatment facility. This results in a reduction of fluoride in the effluent.		Decreases costs of wastewater treatment	N1
***Recycle/Reuse*:** Neutralization of wastewater with soda ash and recycling of wastewater to scrubbers	Aluminum fluoride manufacturing	All pollutants	Neutralizing wastewater with soda ash enables recycling of wastewater in scrubbers because scaling in the equipment is reduced. This in turn reduces the pollutants in and volume of discharges.			N1
***Source Reduction and Recycle/Reuse*:** Recycling of rinsewaters and minimizing product changes	Chromium pigments production	All pollutants	Reusing rinsewater reduces the quantity of wastewater produced. Minimizing product changes requires less equipment clean-up, reduces the amount of rinsewater needed, and reduces the amount of wastewater produced.			N1
***Source Reduction*:** Use of metal anodes instead of graphic anodes	Diaphragm cell process	Lead and toxic organics	Metals anodes reduce the pollutants loads of lead and toxic organics in plant wastewaters		Increases cell power efficiency	N1
***Source Reduction*:** Use of non-contact cooling instead of contact cooling of vapors generated during the concentration of caustic soda	Diaphragm cell process	Caustic soda and salt	Use of non-contact cooling methods reduces the quantity of cooling water used and avoids contamination of cooling water with caustic soda and salt			N1
***Recycle/Reuse*:** Use of ion exchange or reverse osmosis on isolated wastewaters	Chrome pigments production	All pollutants	Use of reverse osmosis or ion exchange removes pollutants from the wastewater, allowing the reuse of the wastewater. This reduces the quantity of wastewater discharged.	Reduces the quantity of sludge produced	Recovers products previously lost in sludges	N1

EXHIBIT 3-11. SUMMARY OF BMPs UTILIZED IN THE INORGANIC CHEMICAL MANUFACTURING INDUSTRY (Continued)

BMP	Targeted Process(es)	Targeted Waste(s)	Benefits to Water	Benefits to Other Media	Other Incentives	Data Sources
Source Reduction: Installation of mechanical scrapers on filters	Copper sulfate and nickel sulfate manufacturing	Copper and nickel	Scrapers on filters eliminate the need for backwashing, which will reduce the quantity of pollutants and the volume of wastewater in the final discharge			N1
Source Reduction: Use of high purity ore in the manufacture of titanium dioxide, an inorganic pigment dye	Synthesis of titanium dioxide	Iron chloride	Reduced ore impurities in the wastewater discharge by using a high quality ore			N2
Recycle/Reuse: Oxidation of ferric chloride used in the manufacture of inorganic pigment dye to recover chlorine	Chlorine recovery	Chlorine	Allows for recovery of chlorine through oxidation, thereby minimizing chlorine discharges			N2

4. RESOURCES AVAILABLE FOR DETERMINING BEST MANAGEMENT PRACTICES

A wide variety of resources can help industry to identify BMPs. Often these resources are the same as those supporting pollution prevention efforts. This chapter contains information about publicly and privately sponsored programs that provide support ranging from onsite assistance to dissemination of information about BMPs. This chapter is based in large part on information contained in an Environmental Protection Agency manual entitled *Pollution Prevention Resources and Training Opportunities in 1992*. Some of the major programs discussed therein are examined in greater detail as part of this manual.

4.1 PURPOSE OF THIS CHAPTER

This chapter provides brief overviews of some of the major sources of information on best management and pollution prevention practices. Most of the resources presented are available through public programs. The organizations that administer these programs may be contacted directly by the user in order to gain access to the program's resources. To assist in this process, this chapter lists each program's address and contact persons as exhibits in each of the subsections on national and international, regional, State, and other programs.

The resources available through these programs vary, with dissemination of fact sheets and case study materials being the predominant form of assistance. The reader should be aware that the assistance provided by some of the programs may be limited. Limitations can include restrictions or the geographical area served, limitations on the subject area, and restrictions on the types of assistance provided. The summary of each program indicates and describes any such restrictions or any costs.

4.2 NATIONAL AND INTERNATIONAL RESOURCES

National and international programs supporting both pollution prevention and the determination of BMPs are accessible to potential users across the United States. The programs discussed in this section include the Pollution Prevention Information Clearinghouse, the International

Cleaner Production Information Clearinghouse, the Waste Reduction Institute for Training and Applications Research, Inc., and the National Technical Information Service. These programs cover a wide variety of topics and provide general information and assistance. Other programs that are limited to a specific region/area of assistance are discussed in Section 4.3.

4.2.1 Pollution Prevention Information Clearinghouse (PPIC)

PPIC is a component of EPA's Pollution Prevention Office program to facilitate pollution prevention information transfer. PPIC is dedicated to reducing or eliminating industrial pollutants through technology transfer, education, and public awareness. PPIC contains technical, financial, programmatic, legislative, and policy information concerning source reduction and recycling efforts in the United States and abroad. PPIC is a free, non-regulatory service of the EPA and is accessible by personal computer, telephone, fax, or mail. The primary components of PPIC are the PPIC repository, the Pollution Prevention Information Exchange System (PIES), and an information hotline. Exhibit 4-1 provides contact information for PPIC.

The PPIC repository is a reference library that includes the most current pollution prevention information in the form of case studies, fact sheets, programmatic and legislative information, and training materials. More than 2,000 documents and reference materials are available through the repository. Information on materials in the repository as well as access to materials in the repository can be obtained by contacting PPIC through the PIES network or the PPIC hotline, both of which are described below. Additionally, interested parties can request information on the contents of the repository by fax or by mail.

PIES is a 24-hour electronic network consisting of a message center, a bulletin board including issue-specific "mini-exchanges," a calendar of events, an online bibliography of materials distributed by PPIC, policy and technical data bases, and a document ordering service. The message center enables users to interact with individual users, EPA, and system operators, or the entire network. Communications can include asking questions, responding to questions, and sharing information and ideas. Some examples of the message center's usefulness include requesting solutions to specific pollution problems, requesting participants in studies, and adding notifications of upcoming events not listed on the calendar of events. EPA also uses the PIES message center to interact with users as part of information exchange efforts.

EXHIBIT 4-1: PPIC CONTACT INFORMATION

CONTACT

Repository Hotline: EPA Headquarters Library

PIES: Rob McCurdy, SAIC

PHONE

EPA: (202) 260-1023

SAIC: (703) 821-4800

FAX

EPA: (202) 260-0178

SAIC: (703) 821-4775

MAIL

Pollution Prevention Information Clearinghouse
EPA Headquarters Library
401 M Street, S.W.
Washington, D.C. 20460

The PIES bulletin board provides information on specific topics. Current bulletins available in the PIES system include PPIC news and announcements, Federal policy statements, Federal grant announcements, feature articles, newsletter updates from national and regional pollution prevention newsletters, and a keyword directory. In addition, several of the regional and State programs (e.g., Northeast Multimedia Pollution Prevention Program, discussed in Section 4.3.1) operate mini-exchanges from the bulletin board system.

The calendar of events contained in PIES lists upcoming training events conducted at the international, national, regional, and State levels. The information contained in the calendar includes topics, dates and times, costs, and contact information for each event. PIES does not provide a registration service; to register for these training events, users must contact training event representatives directly.

The online bibliography contains a list of the materials available in PIES. Generally, the bibliography provides title, author, reference citation, an annotated description, a contact or publication source, and document ordering information. Users can review this material and order the documents through PIES or the hotline automated ordering system.

Policy and technical data bases contained in PIES allow users to obtain online summaries of information by using a keyword search. These data bases include summaries of Federal and State pollution prevention program descriptions, as well as pollution prevention case studies. The Federal and State summaries contain discussions of program objectives and activities, applicable legislation, grants and research projects, and contact information. The case study summaries provide industry profile information followed by a description of pollution prevention program implementation. These descriptions may include materials, chemicals, and feedstocks; initial and final technology descriptions; affected wastes and wastestreams; environmental media involved; costs incurred and costs recovered; and reference information including available facility contact information. Where users wish to gather more detailed information, online summaries can be downloaded directly or ordered using document ordering information provided at the conclusion of each PIES case study summary.

PIES enables the user to access the repository and the document ordering service, and to contact the PPIC technical staff. Instructions for using PIEs appear in Exhibit 4-2. A user who experiences any difficulties entering the PIES system should consult the PIES technical support service, which is part of the information hotline discussed below.

The PPIC hotline is a user-friendly automated voice mail telephone system. Through the automated system, the user can direct any PIES access problems to

EXHIBIT 4-2: INSTRUCTIONS FOR PIES USE

1. Use a computer or dumb terminal equipped with a modem at 1200 or 2400 baud.
2. Set appropriate communications software (e.g., Crosstalk) at 8 data bits, no parity, and 1 stop bit.
3. Call (703) 506-1025 for access.
4. If a user subscribes to SprintNet^{SM}, dial the local access number and enter c20256131 at the @ prompt. (For information on how to subscribe to SprintNet^{SM}, contact the PPIC.)
5. If a user has access to other U.S. private data services that have gateways to SprintNet^{SM}, PIES can be accessed by following the local access procedures established by the data network and typing 311020256131.
6. Once in PIES, follow the prompt commands.

onsite user support personnel. Additionally, the hotline system can provide PIES access to users who do not have access to a computer. In response to hotline requests, technical support personnel will access messages or bulletins and conduct data searches. Finally, the hotline provides a user-friendly automated system to order documents.

Three restrictions are associated with the use of Clearinghouse services:

- Users of the PIES services are limited to 1 hour daily of online access. This ensures that all users of PIES will be granted access.
- Not all case studies summarized in PIES are available through the PPIC document ordering service. Often, the complexity of a document or distribution/copyright restrictions set out by the author or publishing company prevent materials from being available through PPIC. In all cases, however, PPIC users can be referred to the respective author, publishing company, or document distribution center.
- Orders are limited to 10 documents.

4.2.2 International Cleaner Production Information Clearinghouse (ICPIC)

ICPIC is PPIC's sister clearinghouse operated by the United Nations Environment Programme (UNEP). ICPIC provides information to the international community on low- and non-waste producing technologies. ICPIC was developed to coordinate the international exchange of information with an emphasis towards technology transfer to developing countries.

ICPIC has functions and components similar to PPIC, including an electronic information exchange system that is directly accessible to PPIC PIES users. ICPIC contains a message center, bulletins, a calendar of events, case studies, program summaries, an online bibliography, and a directory of contacts. See Exhibit 4-3 for ICPIC access information.

In addition to these components, ICPIC offers a unique resource in that it sponsors working groups that act as forums for exchanging pollution prevention information. Recognized pollution prevention specialists comprise these working groups, which meet regularly to share and gather information on the latest technologies. Working groups have been formed in such industries as textiles, halogenated solvents, leather tanning, biotechnology, and electroplating, and a pulp and paper working group is being formed.

A final unique component of ICPIC is the OzonAction program. OzonAction was established by the United Nations Environment Programme under the Interim Multilateral Ozone Fund (IMOF) of the Montreal Protocol Agreements. OzonAction relays technical and programmatic information on alternatives to all ozone-depleting substances identified by the IMOF. OzonAction provides pollution prevention information to five industry sectors that generate or utilize solvents, coatings and adhesives; halons; aerosols and sterilants; refrigerants; and foams. As part of its technical support, OzonAction provides data bases on solvent substitutes for ozone-depleting substances as compiled by the Industry Cooperative for Ozone Layer Protection. Information contained in OzonAction can be accessed via ICPIC.

EXHIBIT 4-3: ICPIC CONTACT INFORMATION

PHONE

33-1-40-58-88-50

FAX

33-1-40-58-88-74

MAIL

The Director
Industry and Environment Program Activity
United Nations Environment Programme
39-43 Quai André Citroën
75739 Paris CEDEX 15
France

MODEM

1. Use a computer with ASCII capabilities, equipped with a modem and appropriate communications software that operates at 1200 or 2400 baud.
2. Set communications software to 8 data bits, no parity,and 1 stop bit.
3. Dial 33-1-40-58-88-78.
4. Once in ICPIC, follow prompt commands.

4.2.3 Waste Reduction Institute for Training and Applications Research, Inc. (WRITAR)

WRITAR is a private, nonprofit organization designed to identify waste reduction problems, help find their solutions, and facilitate the dissemination of this information to a variety of public and private organizations. WRITAR strives for continual innovation, seeking to provide initiative and direction to other organizations in the field of pollution prevention. The assistance provided by WRITAR are limited in some instances to sponsors. The extent of these limitations can be determined by contacting WRITAR as described in Exhibit 4-4.

EXHIBIT 4-4: WRITAR CONTACT INFORMATION

CONTACT

Terry Foecke or Al Innes

PHONE

(612) 379-5995

FAX

(619) 379-5996

MAIL

WRITAR
1313 5th Street, S.E.
Minneapolis, MN 55414-4502

WRITAR primarily provides pollution prevention assistance for training related activities and policy analyses. Since WRITAR has both public and private roots, it can borrow from an extensive network of knowledgeable individuals who work in private firms, public agencies, and nonprofit organizations to support pollution prevention projects. WRITAR utilizes this expertise and its capabilities to conduct in-depth research to design and deliver pollution prevention training to Federal, State, and local regulators, inspectors, and administrative staff as well as corporate and public audiences. Generally, WRITAR emphasizes the importance of management's approach to successfully implement pollution prevention. WRITAR also conducts industry-specific pollution prevention training for more narrowly defined audiences. Additionally, WRITAR tracks and publishes State legislation that relates to pollution prevention, and analyzes draft legislation and policies for States and localities that are starting their own pollution prevention programs.

4.2.4 National Technical Information Service (NTIS)

NTIS, an agency of the U.S. Department of Commerce, is the central source for the public sale of research, development, engineering, and business reports. The NTIS collection of more than 2 million works covers current industries, business and management studies, foreign and domestic trade, environment and energy, health and the social sciences, translations of foreign reports, general statistics, and many other areas. Approximately 70,000 new technical reports from 200 agencies are added to the NTIS data base annually with nearly one-third of the new additions coming from foreign sources.

The NTIS document service generally is limited to documents created or sponsored by government agencies, including EPA. When government agencies forward reports to NTIS, these items are entered into the NTIS computerized bibliographic database and become part of the archives.

Since an average of 1,300 titles are added to the NTIS collection weekly, NTIS produces a number of printed and electronic awareness services for interested parties. In the environmental field, the NTIS Alert on Environmental Pollution and Control is a twice-monthly bulletin which summarizes recently published environmental-related manuals, reports, and studies. NTIS is a valuable document ordering organization, but does not act as a technical information hotline. Document ordering may be done by mail, phone, or fax. There is a cost associated with each document distributed by NTIS. For the most rapid service, NTIS recommends having the NTIS document number available when ordering documents. Exhibit 4-5 contains NTIS contact information.

4.2.5 Nonpoint Source (NPS) Information Exchange Bulletin Board System (BBS)

The Nonpoint Source (NPS) Information Exchange Bulletin Board System (BBS) provides federal, state and local agencies, private organizations and businesses, and concerned individuals with timely, relevant NPS information, a forum for open discussion, and the ability to exchange computer text and files. The NPS BBS can be used to: read, print, or save to computer disk, current NPS-related articles, reviews and factsheets; exchange computer data including data files, spreadsheets, word processing files and software; post articles and comments on-line for

EXHIBIT 4-5: NTIS CONTACT INFORMATION

CONTACT

National Technical Information Service

PHONE

(800) 336-4700

FAX

(703) 321-8547

MAIL

NTIS
Springfield, VA 22161

others; ask questions and conduct discussions directly with NPS experts; and, exchange private letters and file with others.

To access the NPS BBS, the user will need:

- A PC or terminal
- Telecommunications software (such as Cross Talk or Pro Comm)
- A modem (1200, 2400, or 9600 baud)
- A phone line that can accommodate modem telecommunication

To assist in accessing and using the NPS BBS system, a comprehensive user's guide is available. This guide also describes the various BBS features, and can be obtained by writing to the address shown in Exhibit 4-6.

4.2.6 Office of Water Resource Center

The Office of Water Resource Center is an information clearinghouse for publications and resources available through the U.S. EPA Office of Water. The Resource Center offers:

- Complete Database of Publications
- Document Reference File
- Publication Distribution Management
- NTIS and ERIC Submission Services
- On-Line Ordering for Warehouse
- Data Searches
- Publication Call Referrals
- One Location for Publication Shopping

EXHIBIT 4-6: NPS BBS CONTACT INFORMATION

CONTACT

NPS Information Exchange

PHONE

(301) 589-0205

MAIL

EPA Office of Water
NPS Information Exchange (WH-553)
401 M Street, SW
Washington, DC 20460

This service is available to anyone seeking information regarding U.S. EPA Office of Water publications. (See Exhibit 4-7 for contact information.)

4.3 REGIONAL RESOURCES

EXHIBIT 4-7: OFFICE OF WATER RESOURCE CENTER CONTACT INFORMATION

CONTACT

Office of Water Resource Center

PHONE

(202) 260-7786

HOURS

8:30 am to 5:00 pm

Regional resources are also available for consultation when determining available BMPs and pollution prevention practices. As noted in the following text, many of these resources support limited geographic regions. The regional resources discussed in this section include the Northeast Multimedia Pollution Prevention Program, the Waste Reduction Resource Center for the Southeast, the Pacific Northwest Pollution Prevention Research Center, and EPA's network of pollution prevention contacts and libraries at the regional and headquarters levels.

4.3.1 Northeast Multimedia Pollution Prevention Program (NEMPP)

NEMPP was established in 1989 to help State environmental officials in New England (Connecticut, Maine, Massachusetts, New Hampshire, Rhode Island, and Vermont), New Jersey, and New York implement effective source reduction programs. Technical staff and regulatory officials concerned with air, water, and waste programs participate in the working groups that comprise NEMPP program. The NEMPP program was designed as a resource only to State officials and thus obtaining NEMPP program resources should be coordinated through State officials. (See Exhibit 4-15 in Section 4.4.3 for information on State contacts.)

NEMPP's program provides two components: (1) a clearinghouse of information on pollution prevention/best management practices including technical data, case studies, and a list of pollution prevention experts; and (2) the conduct of training sessions for State officials and industry representatives on source reduction strategies, policies, and technologies.

Additional information on the NEMPP program can be obtained from the NEMPP quarterly newspaper and the NEMPP program mini-exchange established on PIES (see Section 4.2.1 for a description of PIES). To be added to the quarterly newspaper mailing list, the NEMPP program may be contacted as described in Exhibit 4-8. The mini-exchange provides a list of materials in the NEMPP program clearinghouse, a list of experts on pollution prevention in the Northeast, and region-specific articles and newsletters. It also identifies upcoming meetings, conferences, and general source reduction training opportunities.

4.3.2 Waste Reduction Resource Center for the Southeast (WRRC)

WRRC was established in 1988 by the Tennessee Valley Authority and EPA Region 4 to provide multimedia waste reduction support for the States in EPA Region 4: Alabama, Florida, Georgia, Kentucky, Mississippi, South Carolina, North Carolina, and Tennessee. The Center has a collection of technical waste reduction information from Federal Government agencies, all fifty States, and private sources. WRRC consists of several different components including technical assistance by phone and by conducting onsite waste reduction assessments, a repository, an information network, and training.

EXHIBIT 4-8: NEMPP CONTACT INFORMATION

<u>CONTACT</u>

Terri Goldberg

<u>PHONE</u>

(617) 367-8558

<u>FAX</u>

(617) 367-2127

<u>MAIL</u>

Northeast Multimedia Pollution Prevention
Northeast Waste Management Officials Association
85 Merrimac Street
Boston, MA 02114

WRRC provides technical support in developing pollution prevention evaluations, preparing industry-specific reports on waste reduction, and conducting free, onsite non-regulatory site assessments. In these assessments, WRRC personnel draw on their technical experience and knowledge of industrial pollution prevention options.

More than 3,000 pollution prevention documents in the form of journal articles, case studies, technical reports, and books are maintained in the hard copy repository. This includes more than 500 case summaries that describe the application of pollution prevention techniques to many industrial categories and processes. These documents specifically cover economic and technical data, process descriptions, pollution prevention techniques, and implementation strategies, and are often consulted when completing pollution prevention reports.

WRRC also maintains lists of important waste reduction program contacts including persons at EPA Headquarters, EPA Regions, industry trade organizations, universities, other experts, and equipment vendors in the field of pollution prevention for referral and consultation.

While WRRC focuses on supporting groups in EPA Region 4, it offers its assistance to any interested parties. These services are all provided free of charge. WRRC can be contacted as described in Exhibit 4-9.

EXHIBIT 4-9: WRRC CONTACT INFORMATION

CONTACT

Gary Hunt

PHONE

(800) 476-8686

MAIL

Waste Reduction Resource Center for the Southeast
3825 Barrett Drive, Suite 300
Raleigh, NC 27609

4.3.3 Pacific Northwest Pollution Prevention Research Center (PNPPRC)

PNPPRC is a nonprofit public-private partnership dedicated to furthering the goal of multi-media pollution prevention, and to reducing significant waste streams in the Pacific Northwest. PNPPRC is supported through technical assistance grants by industry, environmental and civic organizations, Federal and State governments, and academia. PNPPRC's program is built around the identification of pollution prevention research gaps, the conduct of research, and the communication of research results. PNPPRC activities generally revolve around its pollution prevention grants program.

PNPPRC primarily analyzes technologies and disseminates information based on grant-sponsored research. An information network, another PNPPRC component, connects PNPPRC to associations, industries, small businesses, government, and information sources in the Northwest and to State, regional, and Federal resources. The PNPPRC network educates the public on pollution prevention by providing a pollution prevention research database, a library, in-house publications, and technical and financial referral services. PNPPRC also generates publications and hosts seminars dealing with pollution prevention. PNPPRC performs its services free of charge. However, PNPPRC is limited to providing assistance to facilities located in the States of Alaska, Idaho, Oregon, and Washington, and to the Province of British Columbia. PNPPRC can be accessed using the information provided in Exhibit 4-10.

EXHIBIT 4-10: PNPPRC CONTACT INFORMATION

CONTACT

Madeline Grulich

PHONE

(206) 223-1151

FAX

(206) 223-1165

MAIL

Pacific Northwest Pollution Prevention
Research Center
411 University Street, Suite 1252
Seattle, WA 98101

4.3.4 EPA Offices and Libraries

Each EPA Regional office has identified a contact for pollution prevention. The names and addresses of these contacts are given in Exhibit 4-11. These pollution prevention contacts can provide guidance on programs and associated resources including upcoming regional activities, work group development, industry associations, and a wealth of other references. Such information can be useful to industries seeking assistance in developing a best management or pollution prevention practice plan.

EPA's library system is another resource for those seeking information on pollution prevention and BMPs. Many EPA libraries have specific collections devoted to these areas. These libraries are open to visitors for the conduct of onsite research and frequently allow visitors to generate their own copies for a small copying charge. However, these libraries do not function as photocopying and information dissemination centers. The information contained in Exhibit 4-12 may be used to contact EPA library representatives.

EXHIBIT 4-11. EPA REGIONAL POLLUTION PREVENTION CONTACTS

Contacts: Mark Mahoney and Abby Swaine	**Contact:** Janet Sapadin
Address: U.S. EPA Region I John F. Kennedy Federal Building Boston, MA 02203	**Address:** U.S. EPA Region II 26 Federal Plaza New York, NY 10278
Phone: (215) 597-8327	**Phone:** (212) 264-1925
Contact: Roy Denmark	**Contact:** Carol Monell
Address: U.S. EPA Region III 841 Chestnut Building (3ES43) Philadelphia, PA 19107	**Address:** U.S. EPA Region IV 345 Courtland Street, NE Atlanta, GA 30365
Phone: (215) 597-8327	**Phone:** (404) 347-7109
Contact: Louis Blume	**Contact:** Laura Townsend
Address: U.S. EPA Region V 77 West Jackson Boulevard Chicago, IL 60604-3590	**Address:** U.S. EPA Region VI 1445 Ross Avenue Dallas, TX 75270
Phone: (312) 353-6148	**Phone:** (214) 655-6525
Contact: Alan Wehmeyer	**Contacts:** Don Patton Sharon Childs
Address: U.S. EPA Region VII 726 Minnesota Avenue Kansas City, KS 66101	**Address:** U.S. EPA Region VIII 999 18th Street, Suite 500 Denver, CO 80202-2405
Phone: (913) 551-7336	**Phone:** (303) 293-1456/1471
Contacts: Jesse Baskir Alisa Greene	**Contacts:** Claire Rowlett Carolyn Gangmark
Address: U.S. EPA Region IX 75 Hawthorne Street San Francisco, CA 94105	**Address:** U.S. EPA Region X 1200 Sixth Avenue Seattle, WA 98101
Phone: (415) 744-2038/2189	**Phone:** (206) 553-1099/4072

EXHIBIT 4-12. EPA LIBRARY CONTACT INFORMATION

Contact	Address	Phone
Lois Ramponi, Librarian	Library U.S. EPA Headquarters 401 M Street, S.W. (PM 211A) Washington, DC 20460	(202) 260-3561
Stephen Harmony, Librarian (Contacts)	Center Library Risk Reduction Environmental Laboratory U.S. EPA Headquarters 26 W. Martin Luther King Drive Cincinnati, Ohio 45268	(513) 569-7707
Peg Nelson, Librarian	U.S. EPA Region I John F. Kennedy Federal Building Boston, MA 02203	(617) 565-3300
Eveline Goodman, Librarian	U.S. EPA Region II 26 Federal Plaza New York, NY 10278	(212) 264-2881
Diane M. McCreary, Librarian	U.S. EPA Region III 841 Chestnut Building (3PM 52) Philadelphia, PA 19107	(215) 597-7940
Priscilla Pride, Librarian	U.S. EPA Region IV 345 Courtland Street, NE Atlanta, GA 30365-2401	(404) 347-4216
Ms. Lou W. Tilley, Librarian	U.S. EPA Region V 12th Floor, 77 West Jackson Boulevard Chicago, IL 60604	(312) 886-9506
Linda McGlothlen, Librarian	U.S. EPA Region VI 1445 Ross Avenue, Suite 1200 Dallas, TX 75202-2733	(214) 655-6444
Barbara MacKinnon, Librarian	U.S. EPA Region VII 726 Minnesota Avenue Kansas City, KS 66101	(913) 551-7358
Barbara Wagner, Librarian	U.S. EPA Region VIII 999 18th Street, Suite 500 Denver, CO 80202-2405	(303) 293-1444
Karen Sundheim, Librarian	U.S. EPA Region IX 75 Hawthorne Street, 13th Floor San Francisco, CA 94105	(415) 744-1518
Julienne Sears, Librarian	U.S. EPA Region X 1200 Sixth Avenue Seattle, WA 98101	(206) 553-2969

4.4 STATE, UNIVERSITY, AND OTHER AVAILABLE RESOURCES

Several other available resources can be used to identify applicable pollution prevention and best management practices. These include environmental organization programs, trade association programs, industry-specific programs, and State and university-affiliated programs. This section discusses some of these other resources, including the Center for Waste Reduction Technologies, the Solid Waste Information Clearinghouse, and State and university-affiliated pollution prevention training and information programs.

4.4.1 Center for Waste Reduction Technologies (CWRT)

CWRT was established in 1989 by the American Institute of Chemical Engineers to support industry efforts in meeting the challenge of waste reduction through a partnership between industry, academia, and government. CWRT provides research, education, and information exchange through funding provided by sponsors. CWRT is developing an integrated research program based on the identification of target processes and waste streams and the development of a hierarchy of technologies to address pollutant release reductions or pollutant release elimination.

CWRT is committed to transferring technology and related information to the user community through CWRT-developed how-to publications, training events and conferences, continuing education courses, and links with organizations having related interests. In many cases, a small fee is required for attending training events sponsored by CWRT or for obtaining materials developed and disseminated by CWRT.

CWRT's technology transfer committee works to identify and prioritize waste reduction projects, including BMPs in several technology areas. CWRT's research and development committee targets research to create less polluting technologies such as substitution and process design innovations. CWRT is also developing course materials for graduate and undergraduate engineering curricula and student internship programs as well as continuing education courses for practicing engineers. Contact information for CWRT is provided in Exhibit 4-13.

4.4.2 Solid Waste Information Clearinghouse (SWICH)

SWICH is an information clearinghouse operated by the Solid Waste Association of North America (SWANA) and funded by EPA's Office of Solid Waste, and the Association of Solid Waste Management Professionals. SWICH was developed to help increase the availability of information in the field of solid waste management. SWICH components include an electronic bulletin board (EBB), a library, and a hotline. Contact information for these components is described in Exhibit 4-14.

EXHIBIT 4-13: CWRT CONTACT INFORMATION

CONTACT

Lawrence L. Ross

PHONE

(212) 705-7407

FAX

(212) 752-3297

MAIL

Center for Waste Reduction Technologies
American Institute of Chemical Engineers
345 East 47th Street
New York, NY 10017

EBB functions similar to PIES (see Section 4.2.1 for information on PIES). EBB allows the user to search and order documents from a wide range of solid waste topics including source reduction, recycling, planning, education, public participation, legislation and regulation, waste combustion, composting, collection, waste disposal, and special wastes. EBB also provides updated information on solid waste issues including meeting and conference information, message inquiries, case studies, new technologies, new publications, contact information, and regulatory changes.

Onsite library access is also provided as part of SWICH services, but prior appointments are mandatory. The library contains journals, reports, periodicals, case studies, films, and video tapes, all focusing on solid waste issues. The library also includes a computer work station for access to the EBB. Information in the SWICH library is available for viewing free of charge. There is a per-page charge for photocopying of ordered documents.

Many of the resources available from these programs are free of charge, but not all; the existence of charges should be determined by the user when consulting the program.

4.4.3 State Resources

State programs are another valuable source of information on pollution prevention and best management practices. In addition to grants, technical information, information transfer, and many other integral components of State programs, many States offer training courses and onsite technical assistance to industry either directly or through extension services and academic centers.

The number of State programs prohibits a detailed discussion of each program. Therefore, Exhibit 4-15 presents the formal name of each program and related contact information. Generally, services are limited to facilities located within the State.

EXHIBIT 4-14: SWICH CONTACT INFORMATION

CONTACT

Lori Swan

PHONE

(800) 677-9424

FAX

(301) 585-0297

MAIL

Solid Waste Information Clearinghouse
Solid Waste Association of North America
P.O. Box 7219
Silver Spring, MD 20910

MODEM

1. Use an IBM compatible computer or dumb terminal equipped with a modem and appropriate communications software that operates at 1200 or 2400 baud.

2. Set communications software to 8 data bits, no parity, and 1 stop bit.

3. Dial (301) 585-0204.

4. Once in SWICH, follow prompt commands.

4.4.4 University-Affiliated Resources

University-affiliated resources are primarily centers for research and training in pollution prevention and BMPs that are supported by the particular university and/or industry and State and Federal funds. Since there are too many programs to describe individually, a listing of these programs by State including the contact person is provided in Exhibit 4-16. Many of the resources available from these programs are free of charge, but not all; the existence of charges should be determined by the user when consulting the program.

EXHIBIT 4-15: STATE PROGRAM INFORMATION

STATE	PROGRAM NAME AND SPONSORING AGENCY	CONTACT INFORMATION
Alabama	Alabama Waste Reduction and Technology Transfer (WRATT) Program Alabama Department of Environmental Management	Contact: Daniel E. Cooper Address: 1751 Congressman William L. Dickinson Drive Montgomery, AL 36130 Phone: (205) 271-7939
Alaska	Pollution Prevention Office Alaska Department of Environmental Conservation	Contact: David Wigglesworth Address: P.O. Box O Juneau, AK 99811-1800 Phone: (907) 465-5275
	Waste Reduction Assistance Program (WRAP) and Small Business Hazardous Material Management Project (HMMP) Alaska Health Project	Contact: Kristine Benson Address: 1818 West Northern Lights Boulevard, Suite 103 Anchorage, AK 99517 Phone: (907) 276-2864
Arizona	Arizona Waste Minimization Program Arizona Department of Environmental Quality	Contacts: Stephanie Wilson Dr. J. Andy Soesilo Address: 2005 North Central Avenue Phoenix, AZ 85004 Phone: (602) 257-2318/6995
Arkansas	Arkansas Pollution Prevention Program Arkansas Department of Pollution Control and Ecology	Contact: Robert J. Finn Address: P.O. Box 8913 Little Rock, AR 72219-8913 Phone: (501) 570-2861
	Biomass Resource Recovery Program Arkansas Energy Office	Contact: Ed Davis Address: One State Capital Mall Little Rock, AR 72201 Phone: (501) 682-7322
California	Department of Toxic Substances Control	Contact: Mr. Kim Wilhelm Address: 400 P Street P.O. Box 806 Sacramento, CA 95812-0806 Phone: (916) 324-1807
	Department of Toxic Substances Control Local Government Commission	Contact: Tony Eulo Address: 909 12th Street, Suite 205 Sacramento, CA 95814 Phone: (916) 448-1198
Colorado	Pollution Prevention and Waste Reduction Program Colorado Department of Health	Contact: Neil Kolwey Address: 4210 East 11th Avenue Denver, CO 80220 Phone: (303) 331-4830

EXHIBIT 4-15: STATE PROGRAM INFORMATION

STATE	PROGRAM NAME AND SPONSORING AGENCY	CONTACT INFORMATION
Colorado (Continued)	Pollution Prevention and Waste Reduction Program Colorado Public Interest Research Group (COPIRG)	Contact: Michael Nemecek Address: 1724 Gilpin Street, Denver, CO 80218 Phone: (303) 355-1861
Connecticut	Connecticut Technical Assistance Program (CONNTAP) Connecticut Hazardous Waste Management Service	Contact: Rita Lomasney Address: 900 Asylum Avenue, Suite 360, Hartford, CT 06105-1904 Phone: (202) 241-0777
	Connecticut Department Of Environmental Protection	Contacts: Carmine Di Battista, Elizabeth Flores Address: 165 Capitol Avenue, Hartford, CT 06106 Phone: (203) 566-3437
Delaware	Delaware Pollution Prevention Program Delaware Department of Natural Resources and Environmental Control	Contacts: Philip J. Cherry, Andrea K. Farrell Address: P.O. Box 1401, Kings Highway, Dover, DE 19903 Phone: (302) 739-5071/3822
District Of Columbia	Office Of Recycling D.C. Department of Public Works	Contact: Hampton Cross Address: 65 K Street, Lower Level, Washington, DC 20002 Phone: (202) 939-7116
Florida	Waste Reduction Assistance Program (WRAP) Florida Department of Environmental Regulation	Contact: Janeth A. Campbell Address: 2600 Blair Stone Road, Tallahassee, FL 32399-2400 Phone: (904) 488-0300
Georgia	Georgia Multimedia Source Reduction and Recycling Program Georgia Department of Natural Resources	Contact: Susan Hendricks Address: Floyd Tower East, Suite 1154, 205 Butler Street, S.E., Atlanta, GA 30334 Phone: (404) 656-2833
Hawaii	Hazardous Waste Minimization Program State of Hawaii Department of Health	Contact: Jane Dewell Address: Five Waterfront Plaza, Suite 250, 500 Ala Moana Boulevard, Honolulu, HI 96813 Phone: (808) 586-4226

EXHIBIT 4-15: STATE PROGRAM INFORMATION

STATE	PROGRAM NAME AND SPONSORING AGENCY	CONTACT INFORMATION
Idaho	Division Of Environmental Quality Idaho Department of Health and Welfare	Contacts: Joy Palmer Katie Sewell Address: 1410 North Hilton Street Boise, ID 83720-9000 Phone: (208) 334-5879
Illinois	Office of Pollution Prevention Illinois Environmental Protection Agency	Contacts: Mike Hayes Michael Nechvatal Address: 2200 Churchill Road P.O. Box 19276 Springfield, IL 62794-9276 Phone: (217) 785-0533/8604
Indiana	Office of Pollution Prevention and Technical Assistance Indiana Department of Environmental Management	Contacts: Joanne Joice Charles Sullivan Address: 105 South Meridian Street P.O. Box 6015 Indianapolis, IN 46225 Phone: (317) 232-8172
Iowa	Waste Management Authority Division Iowa Department of Natural Resources	Contacts: Tom Blewett Scott Cahail Address: Wallace State Office Building Des Moines, IA 50319 Phone: (515) 281-8941
Kansas	State Technical Action Plan (STAP) Kansas Department of Health and Environment	Contact: Tom Gross Address: Forbes Field, Building 740 Topeka, KS 66620 Phone: (913) 296-1603
	Kentucky Partners-State Waste Reduction Center	(See Kentucky University Programs)
Louisiana	Louisiana Department of Environmental Quality	Contact: Gary Johnson Address: P.O. Box 82263 Baton Rouge, LA 70884-2263 Phone: (504) 765-0720
Maine	Bureau of Oil and Hazardous Materials Control Maine Department of Environmental Protection	Contact: Scott Whittier Address: State House Station #17 Augusta, ME 04333 Phone: (207) 289-2651
Maryland	Office of Waste Minimization and Recycling Maryland Department of the Environment	Contact: Harry Benson Address: 2500 Broening Highway, Building 40 Baltimore, MD 21224 Phone: (301) 631-3315

EXHIBIT 4-15: STATE PROGRAM INFORMATION

STATE	PROGRAM NAME AND SPONSORING AGENCY	CONTACT INFORMATION
Maryland (Continued)	Maryland Environmental Services	Contact: George G. Perdikakis Address: 2020 Industrial Drive, Annapolis, MD 21401 Phone: (301) 974-7281
Massachusetts	Office of Technical Assistance for Toxics Use Reduction Massachusetts Department of Environmental Management	Contacts: Barbara Kelley, Richard Reibstein Address: 100 Cambridge Street, Boston, MA 02202 Phone: (617) 727-3260
Michigan	Office of Waste Reduction Services Michigan Department of Commerce and Natural Resources	Contact: Larry E. Hartwig Address: 116 West Allegan Street, P.O. Box 30004, Lansing, MI 48909 Phone: (517) 335-1178
Minnesota	Minnesota Pollution Control Agency (MPCA)	Contact: Eric Kilberg Address: Environmental Assessment Office, 520 Lafayette Road, St. Paul, MN 55155 Phone: (612) 296-8643
Mississippi	Mississippi Technical Assistance Program (MISSTAP) and Mississippi Solid Waste Reduction Assistance Program (MISSWRAP)	Contact: Dr. Caroline Hill Address: P.O. Drawer CN, Mississippi State, MS 39762 Phone: (601) 325-8454
	Waste Reduction/Waste Minimization Program Mississippi Department of Environmental Quality	Contact: Thomas E. Whitten Address: P.O. Box 10385, Jackson, MS 39289-0385 Phone: (601) 961-5171
Missouri	Waste Management Program (WMP) Missouri Department of Natural Resources	Contact: Becky Shannon Address: 205 Jefferson Street, P.O. Box 176, Jefferson City, MO 65102 Phone: (314) 751-3176
	Environmental Improvement and Energy Resources Authority (EIERA)	Contacts: Steve Mahfood, Tom Welch Address: 225 Madison Street, P.O. Box 744, Jefferson City, MO 65102 Phone: (314) 751-4919

EXHIBIT 4-15: STATE PROGRAM INFORMATION

STATE	PROGRAM NAME AND SPONSORING AGENCY	CONTACT INFORMATION
Montana	Solid and Hazardous Waste Bureau Montana Department of Health and Environmental Sciences	Contact: Bill Potts Address: Cogswell Building, Helena, MT 59620 Phone: (406) 444-2821
Nebraska	Hazardous Waste Section Nebraska Department of Environmental Control	Contact: Teri Swarts Address: 301 Centennial Mall South, P.O. Box 98922, Lincoln, NE 68509 Phone: (402) 471-4217
Nevada	Nevada Energy Conservation Program Office of Community Services	Contact: Curtis Framel Address: Capitol Complex, 201 South Fall Street, Carson City, NV 89710 Phone: (702) 885-4420
New Hampshire	New Hampshire Pollution Prevention Program New Hampshire Department of Environmental Services	Contact: Vincent R. Perelli Address: 6 Hazen Street, Concord, NH 03301-6509 Phone: (603) 271-2902
New Jersey	New Jersey Office of Pollution Prevention New Jersey Department of Environmental Protection	Contact: Jean Herb (CN-402) Address: 401 East State Street, Trenton, NJ 08625 Phone: (609) 777-0518
	New Jersey Technical Assistance Program (NJTAP) Center for Environmental and Engineering Sciences	Contact: Kevin Gashlin Address: 323 Martin Luther King Boulevard, Newark, NJ 07102 Phone: (201) 596-5864
New Mexico	Municipal Water Pollution Prevention Program New Mexico Environment Department	Contact: Alex Puglisi Address: 1190 St. Francis Drive, P.O. Box 26110, Santa Fe, NM 87502 Phone: (505) 827-2804
New York	Bureau of Pollution Prevention New York State Department of Environmental Conservation	Contact: John Ianotti Address: 50 Wolf Road, Albany, NY 12233-7253 Phone: (518) 457-7276
	New York State Environmental Facilities Corporation New York State Department of Environmental Conservation	Contact: Harold Snow Address: 50 Wolf Road, Albany, NY 12205 Phone: (518) 457-4138

EXHIBIT 4-15: STATE PROGRAM INFORMATION		
STATE	PROGRAM NAME AND SPONSORING AGENCY	CONTACT INFORMATION
New York (Continued)	Erie County Office oOf Pollution Prevention (ECOPP) Erie County Office Building	Contact: Thomas Hersey Address: 95 Franklin Street, Buffalo, NY 14202 Phone: (716) 858-6231
North Carolina	Pollution Prevention Program North Carolina Department of Environment, Health, and Natural Resources	Contact: Gary Hunt Address: P.O. Box 27687, Raleigh, NC 27611-7687 Phone: (919) 571-4100
North Dakota	North Dakota Department of Health and Consolidated Laboratories	Contacts: Neil Knatterud, Teri Lunde Address: P.O. Bos 5520, 1200 Missouri Avenue, Room 302, Bismarck, ND 58502-5520 Phone: (703) 221-5166
Ohio	Ohio Technology Transfer Organization (OTTO) Ohio Department of Development	Contacts: Jeff Shick, Jackie Rudolf Address: 77 South High Street, 26th Floor, Columbus, OH 43255-0330 Phone: (614) 644-4286
	Ohio's Thomas Edison Program	Contact: Dan Berglund Address: 77 South High Street, 26th Floor, Columbus, OH 43215 Phone: (614) 446-3887
	Ohio Environmental Protection Agency	Contacts: Roger Hannahs, Michael W. Kelley, Anthony Sasson Address: P.O. Box 1049, Columbus, OH 43266-0149 Phone: (614) 644-3469
	Ohio Department of Natural Resources Division of Litter Prevention and Recycling	Contact: Helen L. Hurlburt Address: Fountain Square Court, Building F2, Columbus, OH 43224-1387 Phone: (614) 265-6333
Oklahoma	Environmental Health Services Oklahoma State Department of Health	Contacts: Ellen Bussert, Mary Jane Calvey Address: 1000 North East 10th Street, Oklahoma City, OK 73117-1299 Phone: (405) 271-7353

EXHIBIT 4-15: STATE PROGRAM INFORMATION

STATE	PROGRAM NAME AND SPONSORING AGENCY	CONTACT INFORMATION
Oklahoma (Continued)	Pollution Prevention Technical Assistance Program Oklahoma State Department of Health	Contact: Chris Varga Address: 1000 Northeast 10th Street, Oklahoma City, OK 73117-1299 Phone: (405) 271-7047
Oregon	Waste Reduction Assistance Program (WRAP) Oregon Department of Environmental Quality	Contacts: Roy W. Brower, David Rozell, Phil Berry Address: 811 S.W. Sixth Avenue, Portland, OR 97204 Phone: (503) 229-6585
Pennsylvania	Pennsylvania Department of Environmental Resources	Contacts: Keith Kerns, Greg Harder Address: P.O. Box 2063, Harrisburg, PA 17120 Phone: (717) 772-2724
Rhode Island	Hazardous Waste Reduction Program Rhode Island Department of Environmental Management	Contacts: Victor Bell, Richard Enander, Eugene Pepper Address: 83 Park Street, Providence, RI 02903-1037 Phone: (401) 277-3434
South Carolina	Center for Waste Minimization South Carolina Department of Health and Environmental Control	Contact: Jeffrey DeBossone Address: 2600 Bull Street, Columbia, SC 29201 Phone: (803) 734-4715
South Dakota	Waste Management Program South Dakota Department of Environment and Natural Resources	Contact: Vonnie Kallmeyn Address: 319 S. Coteau, c/o 500 E. Capitol Avenue, Pierre, SD 57501 Phone: (605) 773-3153
	Waste Management Program South Dakota Department of Environment and Natural Resources	Contact: Steve Pirner Address: Joe Foss Building, 523 E. Capitol Avenue, Pierre, SD 57501-3181 Phone: (605) 773-3153
Tennessee	Department of Health and Environment	Contact: James Ault Address: 150 9th Avenue, North, Nashville, TN 37219-3657 Phone: (615) 742-6547

EXHIBIT 4-15: STATE PROGRAM INFORMATION

STATE	PROGRAM NAME AND SPONSORING AGENCY	CONTACT INFORMATION
Tennessee (Continued)	Waste Reduction Assessment and Technology Transfer Training Program (WRATT)	Contact: Carol Dugan Address: HB 2G-C, 311 Broad Street, Chattanooga, TN 37406 Phone: (615) 751-4574
	Tennessee Valley Authority	Contact: Steve Hillenbrand Address: Mail Code OCH 2B-K, 602 West Summit Hill Drive, Knoxville, TN 37902 Phone: (615) 632-2101
Texas	Office of Pollution Prevention and Conservation Texas Water Commission	Contacts: Priscilla Seymour, Ph.D., Richard Craig, Robert C. Steckly Address: P.O. Box 13087, Capitol Station, Austin, TX 78711-3087 Phone: (512) 463-7761
Utah	Utah Department of Environmental Quality	Contacts: Rusty Lundberg, Sonja Wallace Address: 288 North 1460 West Street, Salt Lake City, UT 84114-4880 Phone: (801) 538-6170
Vermont	Pollution Prevention Division and Solid Waste Division Vermont Agency of Natural Resources	Contact: Gary Gulka, Paul Markowitz Address: 103 South Main Street, Waterbury, VT 05676 Phone: (802) 244-8702/7831
Virginia	Waste Reduction Assistance Program Virginia Department of Waste Management	Contact: Sharon Kenneally-Baxter Address: Monroe Building, 11th Floor, 101 N. 14th Street, Richmond, VA 23219 Phone: (804) 225-2581
Washington	Waste Reduction, Recycling and Litter Control Program Washington Department of Ecology	Contacts: Stan Springer, Joy St. Germain, Peggy Morgan Address: Mail Stop PV-11, Olympia, WA 98504-8711 Phone: (206) 438-7541
West Virginia	Pollution Prevention and Open Dump Program (PPOD) West Virginia Department of Natural Resources	Contact: Michael Dorsey Address: 1356 Hansford Street, Charleston, WV 25301 Phone: (304) 348-5989

EXHIBIT 4-15: STATE PROGRAM INFORMATION

STATE	PROGRAM NAME AND SPONSORING AGENCY	CONTACT INFORMATION
West Virginia (Continued)	Generator Assistance Program West Virginia Department of Natural Resources	Contacts: Randy Huffman Dale Moncer
		Address: 1356 Hansford Street Charleston, WV 25301
		Phone: (304) 348-4000
Wisconsin	Hazardous Pollution Prevention Audit Grant Program	Contact: Phil Albert
		Address: 123 West Washington Avenue P.O. Box 7979 Madison, WI 53707
		Phone: (608) 266-3075
	Wisconsin Department of Natural Resources	Contact: Lynn Persson Kate Cooper
		Address: P.O. Box 7921 (SW/3) Madison, WI 53707-7921
		Phone: (608) 267-3763
Wyoming	Solid Waste Management Program Wyoming Department of Environmental Quality	Contact: David Finley
		Address: 122 West 25th Street Herschler Building Cheyenne, WY 82002
		Phone: (307) 777-7752

EXHIBIT 4-16. UNIVERSITY AFFILIATED RESOURCES

STATE	PROGRAM NAME AND AFFILIATED UNIVERSITY	CONTACT INFORMATION
Alabama	Hazardous Materials Management and Resource Recovery Program (HAMMARR) University of Alabama	Contact: Dr. Robert Griffin Address: 275 Mineral Industries Building, Box 870203, Tuscaloosa, AL 35487-0203 Phone: (205) 348-8403
California	Environmental Hazards Management Program University of California	Contact: Jon Kindschy Address: University of California Extension, Riverside, CA 92521-0112 Phone: (714) 787-5804
	Center for Waste Reduction Technologies University of California - Los Angeles	Contact: Dr. David Allen Address: University of California at Los Angeles, Los Angeles, CA 90024 Phone: (213) 206-0300
Colorado	Waste Minimization Assessment Center (WMAC) Colorado State University	Contact: Dr. Harry Edwards Address: Mechanical Engineering Department, Fort Collins, CO 80523 Phone: (303) 491-5317
Connecticut	Industrial Environmental Management (IEM) Program Waterbury State Technical College	Contact: Cynthia Donaldson Address: 750 Chase Parkway, Waterbury, CT 06708-3089 Phone: (203) 596-8703/575-8089
Delaware	Delaware Pollution Prevention Program University of Delaware	Contact: Herb Allen Address: Department of Civil Engineering, Newark, DE 19716 Phone: (302) 451-8522/8449
District Of Columbia	The Great Lakes and Mid-Atlantic Hazardous Substance Research Center (GLMA-HSRC) Howard University	Contact: Dr. James H. Johnson, Jr. Address: Department of Civil Engineering, Washington, DC 20059 Phone: (202) 806-6570
Florida	Research Center for Waste Utilization Florida Institute of Technology	Contact: Edwin Korzun Address: 150 West University Boulevard, Melbourne, FL 32901-6988 Phone: (305) 768-8000
	Gulf Coast Hazardous Substance Research Center (GCHSRC) University of Central Florida	See Lamar University, Texas

EXHIBIT 4-16. UNIVERSITY AFFILIATED RESOURCES

STATE	PROGRAM NAME AND AFFILIATED UNIVERSITY	CONTACT INFORMATION
Florida (Continued)	Center for Training, Research, and Education for Environmental Occupations University of Florida	Contact: Dr. James O. Bryant, Jr. Address: 3900 S.W. 63rd Boulevard Gainesville, FL 32608-3848 Phone: (904) 392-9570
Georgia	Environmental Science and Technology Laboratory Georgia Institute of Technology (Georgia Tech)	Contact: Carol Foley Address: Georgia Tech Research Institute Atlanta, GA 30332 Phone: (404) 894-3806
Illinois	Industry Waste Elimination Research Center (IWERC) Illinois Institute of Technology	Contact: Dr. Kenneth E. Noll Address: Pritzker Department of Environmental Engineering IIT Center Chicago, IL 60616 Phone: (312) 567-3536
	Hazardous Waste Research & Information Center (HWRIC) University of Illinois	Contact: Dr. David Thomas Address: One East Hazelwood Drive Champaign, IL 61820 Phone: (217) 333-8940
Indiana	Pollution Prevention Program Purdue University	Contacts: Rick Bossingham, Jeff Burbrink Address: 2129 Civil Engineering Building West Lafayette, IN 47907-1284 Phone: (317) 494-5038
Iowa	Iowa Waste Reduction Center University Of Northern Iowa	Contacts: Kim Gunderson, John Konefes Address: Cedar Falls, IA 50614-0185 Phone: (319) 273-2079
Kansas	Hazardous Substance Research Center (HSRC) Kansas State University	Contact: Dr. Larry E. Erickson Address: Durland Hall, Room 105 Manhattan, KS 66506-5102 Phone: (913) 532-5584
	Center for Environmental Education and Training University of Kansas	Contact: Lani Heimgardner Address: 6330 College Boulevard Overland Park, KS 66211 Phone: (913) 491-0810
Kentucky	Kentucky Partners - State Waste Reduction Center University of Louisville	Contact: Joyce St. Clair Address: Ernst Hall, Room 312 Louisville, KY 40292 Phone: (502) 588-7260

EXHIBIT 4-16. UNIVERSITY AFFILIATED RESOURCES

STATE	PROGRAM NAME AND AFFILIATED UNIVERSITY	CONTACT INFORMATION
Kentucky (Continued)	Waste Minimization Assessment Center University of Louisville	Contact: Marvin Fleischman Address: Department of Chemical Engineering, Louisville, KY 40292 Phone: (502) 588-6357
Louisiana	Hazardous Waste Research Center (HWRC) Louisiana State University	Contact: David Constant Address: 3418 CEBA Building, Baton Rouge, LA 70803 Phone: (504) 388-6770
	Gulf Coast Hazardous Substance Research Center (GCHSRC) Louisiana State University	See Lamar University, Texas
	Technical Extension Service University of Maryland	Contact: Travis Walton Address: Engineering Research Center, College Park, MD 20742 Phone: (301) 454-1941
Massachusetts	Center for Technology, Policy and Industrial Development Massachusetts Institute of Technology	Contact: John Enhrenfeld Address: E40-241, Cambridge, MA 02139 Phone: (617) 253-7753
	Tufts Environmental Literacy Institute (TELI) Tufts University	Contact: Dr. Anthony Cortese Address: Office of Environmental Programs, 474 Boston Avenue, Curtis Hall, Medford, MA 02155 Phone: (617) 627-3452
	The Center for Environmental Management Tufts University	Contact: Dr. William R. Moomaw Address: 474 Boston Avenue, Curtis Hall, Medford, MA 02155 Phone: (617) 381-3486
	Toxics Use Reduction Institute University of Lowell	Contact: Jack Luskin Address: 1 University Avenue, Lowell, MA 01852 Phone: (508) 934-3275
Michigan	Waste Reduction and Management Program (WRMP) Grand Valley State University School of Engineering	Contact: Dr. Paul Johnson Address: 301 W. Fulton, Room 617, Grand Rapids, MI 49504 Phone: (616) 771-6750

EXHIBIT 4-16. UNIVERSITY AFFILIATED RESOURCES

STATE	PROGRAM NAME AND AFFILIATED UNIVERSITY	CONTACT INFORMATION
Michigan (Continued)	Environmental Engineering Center For Waste And Waste Management Michigan Technological University	Contact: Neil Hutzler Address: Environmental Engineering Center, 1400 Townsend Drive, Houghton, MI 49931 Phone: (906) 487-2098
	Pollution Prevention Center for Curriculum Development and Dissemination University of Michigan	Contact: Dr. Gregory A. Keoleian Address: School of Natural Resources, Dana Building, 430 E. University, Ann Arbor, MI 48109-1115 Phone: (313) 764-1412
	The Great Lakes And Mid-Atlantic Hazardous Substance Research Center (GLMA-HSRC) University of Michigan	Contact: Dr. Walter Weber Address: Suite 181 Engineering 1-A, Ann Arbor, MI 48109-2125 Phone: (313) 763-2274
Minnesota	Minnesota Technical Assistance Program University of Minnesota	Contact: David Simmons, Cindy McCombs Address: 1315 5th Street, S.E., Suite 207, Minneapolis, MN 55414 Phone: (612) 627-4646
Mississippi	Mississippi Technical Assistance Program And Mississippi Solid Waste Reduction Assistance Program Mississippi State University	Contacts: Dr. Don Hill, Dr. Caroline Hill, Dr. June Carpenter Address: P.O. Drawer CN, Mississippi State, MS 39762 Phone: (601) 325-8454
	Gulf Coast Hazardous Substance Research Center (GCHSRC) Mississippi State University	See Lamar University, Texas
Nevada	Nevada Small Business Development Center University of Nevada - Reno	Contact: Kevin Dick Address: Reno, NV 89557-0100 Phone: (702) 784-1717
New Jersey	Hazardous Substance Management Research Center New Jersey Institute of Technology	Contact: Dr. Kevin Gashlin Address; Advanced Technology Center Building, 323 Martin Luther King Boulevard, University Heights, Newark, NJ 07102 Phone: (201) 596-5864
New Mexico	Waste-Management Education And Research Consortium (WERC) New Mexico State University	Contact: John S. Townsend Address: Box 30001, Department 3805, Las Cruces, NM 88003-0001 Phone: (505) 646-2038

EXHIBIT 4-16. UNIVERSITY AFFILIATED RESOURCES

STATE	PROGRAM NAME AND AFFILIATED UNIVERSITY	CONTACT INFORMATION
New York	Hazardous Waste and Toxic Substance Research and Management Center Clarkson University	Contact: Thomas L. Theis Address: Rowley Laboratories, Potsdam, NY 13699 Phone: (315) 268-6542
	Waste Management Institute Cornell University	Contact: Richard Schuler Address: 313 Hollister Hall, Ithaca, NY 14853 Phone: (607) 255-8674
North Carolina	EPA Research Center for Waste Minimization and Management North Carolina State University	Contacts: Dr. Michael Overcash, Dr. Cliff Kaufman Address: Box 7905, Raleigh, NC 27695-2325 Phone: (919) 515-2325
	EPA Research Center for Waste Minimization and Management University of North Carolina - Chapel Hill	Contact: Dr. William H. Glaze Address: Department of Environmental Science & Engineering, Chapel Hill, NC 27514 Phone: (919) 966-1024
Ohio	American Institute For Pollution Prevention (AIPP) University Of Cincinnati	Contact: Jean Boddocsi Address: Office of the University Dean for Research, Cincinnati, OH 45221 Phone: (513) 556-4532
	RCRA Generator Training Program University Of Findlay	Contact: George Kleevic Address: P.O. Box 538, St. Clairsville, OH 43950 Phone: (614) 695-5036
Oregon	Waste Reduction Assistance Program Oregon State University	Contact: Dr. Ken Williamson Address: Civil Engineering Department, Apperson 206, Corvallis, OR 97331-2302 Phone: (503) 754-2751
Pennsylvania	Pennsylvania Technical Assistance Program (PENNTAP) Pennsylvania State University	Contact: Jack Gido Address: 248 Calder Way, Suite 306, University Park, PA 16801 Phone: (814) 865-1914
	National Technology Applications Corporation University of Pittsburgh (NETAC)	Contact: Devon Streit Address: Applied Research Center, 615 William Pitt Way, Pittsburgh, PA 15238 Phone: (412) 826-5511

EXHIBIT 4-16. UNIVERSITY AFFILIATED RESOURCES

STATE	PROGRAM NAME AND AFFILIATED UNIVERSITY	CONTACT INFORMATION
Pennsylvania (Continued)	Center For Hazardous Materials Research (CHMR) University oOf Pittsburgh	Contacts: Dr. Edgar Berkey, Roger Price Address: 320 William Pitt Way, Pittsburgh, PA 15238 Phone: (412) 826-5320 or (800) 334-CHMR
Rhode Island	Chemical Engineering Department University of Rhode Island	Contact: Prof. Stanley M. Barnett Address: Crawford Hall, Kingston, RI 02881 Phone: (401) 792-2443
South Carolina	Hazardous Waste Management Research Fund Clemson University	Contact: Eric Snider, PhD., P.E. Address: Continuing Engineering Education, P.O. Drawer 1607, Clemson, SC 29633 Phone: (803) 656-3308
	Hazardous Waste Management Research Fund University of South Carolina	Contact: Doug Dobson Address: Institute of Public Affairs, Gambrell Hall, 4th Floor, Columbia, SC 29208 Phone: (803) 777-8157
Tennessee	Waste Reduction Assessment and Technology Transfer Training Program University of Tennessee	Contact: Cam Metcalf Address: Center for Industrial Services, 226 Capitol Boulevard Building, Suite 606 Nashville, TN 37219 Phone: (615) 242-2456
	Waste Minimization Assessment Center University of Tennessee	Contact: Dr. Richard J. Jendrucko Address: Department of Engineering, Science and Mechanics, 310 Perkins Hall, Knoxville, TN 37996-2030 Phone: (615) 974-7682
Texas	EPA Research Center for Waste Minimization and Management Texas A & M University	Contact: Dr. Kirk Brown Address: Department of Soil and Crop Science, College Station, TX 77843 Phone: (409) 845-5251
	Gulf Coast Hazardous Substance Research Center (GCHSRC) Lamar University	Contact: Dr. William Cawley Address: P.O. Box 10613, Beaumont, TX 77710 Phone: (409) 880-8707
	Gulf Coast Hazardous Substance Research Center (GCHSRC) University of Houston	See Lamar University, Texas

EXHIBIT 4-16. UNIVERSITY AFFILIATED RESOURCES

STATE	PROGRAM NAME AND AFFILIATED UNIVERSITY	CONTACT INFORMATION
Texas (Continued)	Gulf Coast Hazardous Substance Research Center (GCHSRC) University of Texas - Austin	See Lamar University, Texas
	Center For Environmental Technologies Texas Tech University	Contact: Dr. John R. Bradford Address: P.O. Box 43121, Lubbock, TX 79409-3121 Phone: (806) 742-1413
	Environmental Institute for Technology Transfer (EITT) University Of Texas - Arlington	Contacts: Dr. Gerald I. Nehman, Dr. Victorio Argento Address: Box 19050, Arlington, TX 76019 Phone: (817) 273-2300
Utah	Department of Chemical Engineering University of Utah	Contact: JoAnn S. Lighty Address: 3290 MEB, Salt Lake City, UT 84112 Phone: (801) 581-5763
	Safety Office Utah State University	Contact: Nancy Fox Address: UMC 8318, Logan, UT 84322-8315 Phone: (801) 750-2752
Wisconsin	Engineering Professional Development Program University oOf Wisconsin - Madison	Contact: Pat Eagan Address: College of Engineering, 432 North Lake Street, Madison, WI 53706 Phone: (608) 263-7429
	Solid aAnd Hazardous Waste Education Center University of Wisconsin	Contact: David Liebel, Wayne Pferdehirt Address: 529 Lowell Hall, 610 Langdon Street, Madison, WI 53703 Phone: (608) 265-2360

APPENDIX A — BEST MANAGEMENT PRACTICES PLAN DEVELOPMENT CHECKLIST

THE BMP COMMITTEE	
I. Has a best management practices (BMP) committee been created to develop the BMP plan?	☐ yes ☐ no ☐ n/a
II. Is a discussion of the BMP committee provided in the BMP plan?	☐ yes ☐ no ☐ n/a
A. Is there a complete list of persons chosen to serve on the BMP committee in the BMP plan?	☐ yes ☐ no ☐ n/a
B. Are backup people listed with phone numbers?	☐ yes ☐ no ☐ n/a
C. Are other individuals, non-committee members, available for technical input when necessary?	☐ yes ☐ no ☐ n/a
III. Has development of the BMP committee as part of an existing committee performing similar functions been considered to reduce duplicity of efforts?	☐ yes ☐ no ☐ n/a
IV. Have personnel selections and responsibility designations been determined to ensure the committee's effective function?	☐ yes ☐ no ☐ n/a
A. Has a lead committee member been chosen to chair the BMP committee?	☐ yes ☐ no ☐ n/a
1. Do qualifications of the lead committee member include managing large projects?	☐ yes ☐ no ☐ n/a
2. Is the lead committee member highly motivated to develop and implement the BMP plan?	☐ yes ☐ no ☐ n/a
3. Is the lead committee member familiar with all committee members and their area of expertise?	☐ yes ☐ no ☐ n/a
B. Is the BMP committee made up of company personnel that are knowledgeable in the BMP areas of concern?	☐ yes ☐ no ☐ n/a
1. Do employees of sufficient expertise comprise the BMP committee?	☐ yes ☐ no ☐ n/a
2. Are the BMP committee members familiar with all pertinent Federal, state and local regulations?	☐ yes ☐ no ☐ n/a
C. Do committee members include individuals from the company structure that are in decision-making positions so that the committee will be able to make decisions without spending time waiting for approvals?	☐ yes ☐ no ☐ n/a

THE BMP COMMITTEE (Continued)	
D. Are employee interests represented by the BMP committee representatives (e.g., have union stewards, equal opportunity representatives, operations or maintenance directors, technical specialists, and knowledgeable employees been considered as potential committee members)?	☐ yes ☐ no ☐ n/a
E. Does the size of the committee reflect the size and complexity of the facility?	☐ yes ☐ no ☐ n/a
1. Is the committee small enough so that communication is open and interactive, yet large enough to allow relevant input from all sectors of the facility?	☐ yes ☐ no ☐ n/a
2. Has the committee been organized such that consensus on BMP issues can be easily reached?	☐ yes ☐ no ☐ n/a
3. Have personality and interaction considerations been taken into account when selecting the BMP committee?	☐ yes ☐ no ☐ n/a
F. Are the roles and responsibilities of each BMP committee member clearly defined?	☐ yes ☐ no ☐ n/a
V. Do BMP committee personnel have time available to meet their responsibilities?	☐ yes ☐ no ☐ n/a
VI. Are coordination procedures set forth for the effective function of the BMP committee?	☐ yes ☐ no ☐ n/a
A. Are weekly memos considered to keep personnel informed?	☐ yes ☐ no ☐ n/a
B. Are meetings planned that BMP personnel from all shifts can attend?	☐ yes ☐ no ☐ n/a

THE BMP POLICY STATEMENT	
I. Is there a BMP policy statement included in the BMP plan?	☐ yes ☐ no ☐ n/a
II. Does the BMP policy statement clearly and coherently state the objectives for control of toxic and hazardous substances and indicate management's support?	☐ yes ☐ no ☐ n/a
A. Is it signed by the facility management or corporate officers?	☐ yes ☐ no ☐ n/a

THE BMP POLICY STATEMENT (Continued)	
B. Is the policy statement written in a positive tone that emphasizes the advantages of the BMP plan with respect to pollution prevention at the facility?	☐ yes ☐ no ☐ n/a
1. Are the savings in raw materials, pollution control and liability costs mentioned?	☐ yes ☐ no ☐ n/a
2. Is enhanced safety of the work environment emphasized?	☐ yes ☐ no ☐ n/a
3. Is increased production efficiency mentioned?	☐ yes ☐ no ☐ n/a
III. Is there a method to address employee suggestions or concerns pertaining to the BMP policy?	☐ yes ☐ no ☐ n/a
A. Have training sessions or seminars been developed to explain the BMP policy statement?	☐ yes ☐ no ☐ n/a
B. Is there a point of contact listed on the BMP policy statement for employee questions and comments?	☐ yes ☐ no ☐ n/a
IV. Have provisions been made to keep employees aware of the BMP policy?	☐ yes ☐ no ☐ n/a
A. Was a memo distributed to all employees affected by the policy statement describing the new procedures that will be instituted?	☐ yes ☐ no ☐ n/a
B. Is the BMP policy statement posted in key locations where many employees will read it?	☐ yes ☐ no ☐ n/a
RELEASE IDENTIFICATION AND ASSESSMENT	
I. Has a release identification and assessment been performed to determine the need for specific BMPs and to ascertain areas to which the BMP plan should focus?	☐ yes ☐ no ☐ n/a
II. Are details of the release identification and assessment summarized in the BMP plan?	☐ yes ☐ no ☐ n/a
A. Has existing information been reviewed for relevance to release identification and assessment (e.g., spill prevention control and countermeasure plans; preparedness, prevention, and contingency plans; storm water pollution prevention plans; and the NPDES permit application)?	☐ yes ☐ no ☐ n/a
B. Have pollutant sources been adequately identified?	☐ yes ☐ no ☐ n/a

RELEASE IDENTIFICATION AND ASSESSMENT (Continued)	
1. Has a standard form been developed for recording currently and potentially released substances on the site?	☐ yes ☐ no ☐ n/a
a. Are areas noted in which the chemicals are present or may be present?	☐ yes ☐ no ☐ n/a
i. Are currently released amounts noted?	☐ yes ☐ no ☐ n/a
ii. Are amounts of materials which may be potentially released also noted?	☐ yes ☐ no ☐ n/a
b. Where potential for release exists, are methods of control for materials also noted (i.e., dikes, pumps)?	☐ yes ☐ no ☐ n/a
i. Are controls for potential releases adequate to discount consideration?	☐ yes ☐ no ☐ n/a
ii. Have factors been assigned based on the confidence of those measures in controlling releases?	☐ yes ☐ no ☐ n/a
2. Have maps and drawings been used to describe the facility?	☐ yes ☐ no ☐ n/a
a. Are outfall locations and drainage patterns shown on these maps?	☐ yes ☐ no ☐ n/a
i. Is the contour of the land considered in predicting the direction of flow?	☐ yes ☐ no ☐ n/a
ii. Is the receiving water in the area clearly shown and named on the maps?	☐ yes ☐ no ☐ n/a
b. Are the plant features (i.e., locations of materials and pollutant controls) clearly marked on the site map?	☐ yes ☐ no ☐ n/a
3. Are the locations of outfalls, chemicals and plant features clearly marked on the site map and referenced?	☐ yes ☐ no ☐ n/a
4. Has an inspection been conducted to verify information gathered through data review?	☐ yes ☐ no ☐ n/a
C. Have currently and potentially discharged pollutants been prioritized based on the amount of discharge and their hazards to human health and the environment?	☐ yes ☐ no ☐ n/a

RELEASE IDENTIFICATION AND ASSESSMENT (Continued)	
1. Has the Material Safety Data Sheet for the potential pollutants been reviewed?	☐ yes ☐ no ☐ n/a
2. Have sources of available exposure limits to protect human health been reviewed (e.g., NIOSH Handbook, and the handbook of industrial standards from ACGIH)?	☐ yes ☐ no ☐ n/a
3. Have health and safety personnel been consulted for an accurate assessment of potential health risks?	☐ yes ☐ no ☐ n/a
D. Have pathways been identified through which pollutants found at an area/site might reach environmental and human receptors?	☐ yes ☐ no ☐ n/a
1. Has the materials inventory been examined in combination with areas of actual or potential release to identify release mechanisms and receptor media?	☐ yes ☐ no ☐ n/a
2. Has a description of releases and pathways for each pollutant source been prepared?	☐ yes ☐ no ☐ n/a
3. Have all logical alternative pathways been considered?	☐ yes ☐ no ☐ n/a
E. Have both actual and potential releases been prioritized?	☐ yes ☐ no ☐ n/a
1. Has information about the release been combined with information about the toxicity or hazards associated with each pollutant found at the facility?	☐ yes ☐ no ☐ n/a
2. Have actual and potential releases been ranked as to high, medium, or low probability for both current and potential releases?	☐ yes ☐ no ☐ n/a
III. Has the use of non-company representatives with expertise in conducting a release identification and assessment been considered?	☐ yes ☐ no ☐ n/a
IV. Is a release identification and assessment conducted prior to the implementation of a new process or the use of a new material?	☐ yes ☐ no ☐ n/a
V. Has materials compatibility been considered in this evaluation?	☐ yes ☐ no ☐ n/a

RELEASE IDENTIFICATION AND ASSESSMENT (Continued)	
A. Have assessments been made of materials which may be potentially released at the same outfall/drainage point?	☐ yes ☐ no ☐ n/a
B. Have assessments been made as to the compatibility between containers and materials when determining confidence ratings of controls?	☐ yes ☐ no ☐ n/a
GOOD HOUSEKEEPING	
I. Is good housekeeping addressed in the BMP plan?	☐ yes ☐ no ☐ n/a
II. Is the good housekeeping program focused on the areas identified during the release identification and assessment stage as having the highest potential for environmental releases?	☐ yes ☐ no ☐ n/a
III. Can the good housekeeping program as part of the BMP plan be incorporated into the existing programs or standard operating procedures?	☐ yes ☐ no ☐ n/a
IV. Will the good housekeeping program adequately address releases resulting from poor housekeeping?	☐ yes ☐ no ☐ n/a
A. Is the facility well organized?	☐ yes ☐ no ☐ n/a
1. Are all packaged and bagged chemicals properly stored in appropriate storage areas?	☐ yes ☐ no ☐ n/a
a. Are containers, drums, and bags stored away from direct traffic routes to prevent accidental spills?	☐ yes ☐ no ☐ n/a
b. Has storing containers on pallets or similar devices to prevent corrosion of containers been considered?	☐ yes ☐ no ☐ n/a
c. Are containers stacked according to manufacturers' instructions to avoid damaging them from improper weight distribution?	☐ yes ☐ no ☐ n/a
2. Have all containers been labelled to show the name and type of substance, stock number, expiration date, health hazards, suggestions for handling, and first aid information?	☐ yes ☐ no ☐ n/a
3. Has alleviation of space constraints been considered to help implement good housekeeping?	☐ yes ☐ no ☐ n/a

GOOD HOUSEKEEPING (Continued)	
a. Are walkways and passageways easily accessible, safe, and free of protruding objects and equipment?	☐ yes ☐ no ☐ n/a
B. Is the facility maintained in a clean fashion?	☐ yes ☐ no ☐ n/a
1. Is there any evidence of drippings from equipment or machinery?	☐ yes ☐ no ☐ n/a
2. Is there evidence of dust in the air or on the floor?	☐ yes ☐ no ☐ n/a
C. Are released materials regularly and easily mitigated?	☐ yes ☐ no ☐ n/a
1. Are recycle and waste disposal areas located close to waste generation areas to prevent inappropriate disposal of waste?	☐ yes ☐ no ☐ n/a
a. Where disposal is the current option, have recycle measures been considered?	☐ yes ☐ no ☐ n/a
b. Is the material involved in small incidents recovered, rather than cleaned/flushed with water?	☐ yes ☐ no ☐ n/a
2. Are physical and mechanical cleanup equipment readily available and properly stored away in appropriate locations?	☐ yes ☐ no ☐ n/a
3. Are floors and ground surfaces kept dry and clean by using brooms, shovels, vacuum cleaners, or cleaning machines?	☐ yes ☐ no ☐ n/a
D. Have procedures been developed to maintain good housekeeping measures and ensure that all materials and equipment be returned or replaced in their designated areas?	☐ yes ☐ no ☐ n/a
1. Are all employees aware of the importance of good housekeeping through training?	☐ yes ☐ no ☐ n/a
2. Are publicity posters, bulletin boards, and employee publications used for good housekeeping programs?	☐ yes ☐ no ☐ n/a
3. Are written instructions distributed detailing good housekeeping procedures?	☐ yes ☐ no ☐ n/a

GOOD HOUSEKEEPING (Continued)	
4. Do shift supervisors and other personnel in positions of authority uphold the good housekeeping procedures to demonstrate by example to their employees?	☐ yes ☐ no ☐ n/a
E. Are there regular housekeeping inspections to check for good housekeeping problems?	☐ yes ☐ no ☐ n/a
IV. Have materials compatibility issues been considered in the storage, clean-up, recycle/reuse, and disposal aspects of the good housekeeping program?	☐ yes ☐ no ☐ n/a
PREVENTIVE MAINTENANCE	
I. Are preventive maintenance (PM) procedures covered in the BMP plan?	☐ yes ☐ no ☐ n/a
II. Is the PM program focused on the areas identified during the release identification and assessment stage as having the highest potential for environmental releases?	☐ yes ☐ no ☐ n/a
III. Can the PM program as part of the BMP plan be incorporated into the existing PM program?	☐ yes ☐ no ☐ n/a
IV. Is the PM plan adequate to prevent environmental releases resulting form poor maintenance activities?	☐ yes ☐ no ☐ n/a
A. Has an equipment inventory system been set up?	☐ yes ☐ no ☐ n/a
1. Does it provide the equipment location and identification information?	☐ yes ☐ no ☐ n/a
2. Has equipment been labelled with assigned names or numbers and identification information?	☐ yes ☐ no ☐ n/a
3. Are index cards, prepared forms or checklists, or computer programs used to record inventory information?	☐ yes ☐ no ☐ n/a
4. Is an inventory kept of all maintenance materials needed?	☐ yes ☐ no ☐ n/a
a. Does this include the materials/parts description, number, item specifications, ordering information, vendor addresses and phone numbers, storage locations, maximum order quantities, and costs?	☐ yes ☐ no ☐ n/a

PREVENTIVE MAINTENANCE (Continued)	
b. Has a system been considered to track the items needed to make simple repairs, parts that are vulnerable to breakage, and parts with long delivery times or that are difficult to obtain?	☐ yes ☐ no ☐ n/a
c. Is the stockpile of spare parts adequate?	☐ yes ☐ no ☐ n/a
i. Does it include parts that are hard to obtain?	☐ yes ☐ no ☐ n/a
ii. Does it include specialized tools?	☐ yes ☐ no ☐ n/a
B. Have preventive maintenance requirements been determined (e.g., recommended schedules and specifications for lubrication, parts replacement, equipment testing, and/or maintenance of spare parts)?	☐ yes ☐ no ☐ n/a
1. Have manufacturer's references, pamphlets, and guidebooks been consulted?	☐ yes ☐ no ☐ n/a
2. Have maintenance schedules and specifications for all equipment been summarized for easy review and understanding (i.e., in tabular form or on index cards for each piece of equipment)?	☐ yes ☐ no ☐ n/a
3. Are maintenance personnel aware of the schedules and specifications (i.e., by placing on a blackboard)?	☐ yes ☐ no ☐ n/a
C. Are PM activities conducted at frequencies and specifications at least as stringent as the manufacturer's recommendations?	☐ yes ☐ no ☐ n/a
1. Are methods set forth to ensure that maintenance activities have been conducted?	☐ yes ☐ no ☐ n/a
a. Are there records of preventive maintenance schedules, tests, inspections, repairs, lubrications, etc.?	☐ yes ☐ no ☐ n/a
b. Are periodic inspections conducted to determine if schedules are being met or work is being completed?	☐ yes ☐ no ☐ n/a
2. Do PM activities include the periodic testing for structural soundness (e.g., making sure that storage tanks are solid and strong enough to hold materials)?	☐ yes ☐ no ☐ n/a

PREVENTIVE MAINTENANCE (Continued)	
3. Has special attention been given to equipment that frequently breaks down (i.e., more frequent PM, making sure that spare parts are always in supply)?	☐ yes ☐ no ☐ n/a
D. Has a system been developed for keeping records of PM activities?	☐ yes ☐ no ☐ n/a
1. Has a tracking system been developed to monitor upkeep PM activities and cost?	☐ yes ☐ no ☐ n/a
a. Can the preventive maintenance records be used in determining whether equipment should be repaired or replaced?	☐ yes ☐ no ☐ n/a
b. Has a replacement program been considered for equipment, including vessels and tanks, based on age and shape of equipment?	☐ yes ☐ no ☐ n/a
2. Is the tracking system well organized and easy to use?	☐ yes ☐ no ☐ n/a
a. Can specifications and schedules for PM be easily recognized?	☐ yes ☐ no ☐ n/a
b. Can PM activities be easily verified based on records?	☐ yes ☐ no ☐ n/a
V. Is there down-time associated with conducting PM?	☐ yes ☐ no ☐ n/a
A. Are PM activities requiring down-time scheduled to prevent disruptions to plant operations?	☐ yes ☐ no ☐ n/a
B. Are individuals affected by the down-time notified in advance?	☐ yes ☐ no ☐ n/a

INSPECTIONS	
I. Is an inspection program addressed in the BMP plan?	☐ yes ☐ no ☐ n/a
II. Can the inspection program, as part of the BMP plan, be incorporated into current standard operation procedures?	☐ yes ☐ no ☐ n/a
III. Is the inspection program focused on the areas identified during the release identification and assessment stage as having the highest potential for environmental releases?	☐ yes ☐ no ☐ n/a
IV. Is the inspection program adequate such that current and potential releases are controlled?	☐ yes ☐ no ☐ n/a

INSPECTIONS (Continued)	
A. Does the inspection program include the identification of different inspections with various scopes (e.g., security scan, walk-through, site review, BMP plan oversight inspection, and BMP plan re-evaluation inspection)?	☐ yes ☐ no ☐ n/a
B. Is a schedule developed for conducting inspections?	☐ yes ☐ no ☐ n/a
1. Is a comprehensive inspection performed at least once per year?	☐ yes ☐ no ☐ n/a
2. Are inspection designed to overlap to provide oversight mechanisms?	☐ yes ☐ no ☐ n/a
3. Are inspection conducted more frequently in areas of highest concern?	☐ yes ☐ no ☐ n/a
4. Are inspectors alternated and/or is a team approach used during inspections to conduct a more thorough review?	☐ yes ☐ no ☐ n/a
C. Are inspections conducted by qualified personnel (e.g., security, technical personnel, supervisors)?	☐ yes ☐ no ☐ n/a
1. Are areas reviewed for evidence of pollutants releases (i.e., spills, discolorations, odor)?	☐ yes ☐ no ☐ n/a
2. Are areas of concern inspected with greater intensity?	☐ yes ☐ no ☐ n/a
3. Are site maps used to ensure that potentially released materials and drainage patterns are evaluated?	☐ yes ☐ no ☐ n/a
4. Is a list of personnel responsible for inspections indicated in the BMP plan (i.e, foreman, area supervisor, department manager, safety coordinator, environmental control coordinator) provided?	☐ yes ☐ no ☐ n/a
D. Has an inspection form been developed including a space for a narrative report and/or checklist of areas to inspect?	☐ yes ☐ no ☐ n/a
1. Are checklists available for each type of inspection, as necessary?	☐ yes ☐ no ☐ n/a
2. Are narrative discussions provided with each checklist, as necessary?	☐ yes ☐ no ☐ n/a

INSPECTIONS (Continued)	
E. Are reports prepared summarizing inspection results and detailing follow-up actions?	☐ yes ☐ no ☐ n/a
1. Is some method of reporting each inspection provided (i.e., verbal notification, written report)?	☐ yes ☐ no ☐ n/a
2. Are reports reviewed by designated personnel for the necessity for quick response or remedial action?	☐ yes ☐ no ☐ n/a
V. Are employees encouraged to periodically conduct informal visual inspections?	☐ yes ☐ no ☐ n/a
VII Has the use of non-regulatory support from EPA, States, or universities when conducting inspections, particularly the BMP plan oversight and evaluation/re-evaluation inspections, been considered?	☐ yes ☐ no ☐ n/a
SECURITY	
I. Is a security plan included in the BMP plan?	☐ yes ☐ no ☐ n/a
II. Can the inspection program as part of the BMP plan be incorporated into current standard operation procedures?	☐ yes ☐ no ☐ n/a
III. Does the security program, as part of the BMP plan, focus on areas identified during the release identification and assessment stage as having the highest potential for harmful environmental releases?	☐ yes ☐ no ☐ n/a
IV. Is the security program adequate such that current and potential releases are controlled?	☐ yes ☐ no ☐ n/a
A. Have the security personnel been considered for use in the conduct of visual inspections to identify actual or potential releases of concern?	☐ yes ☐ no ☐ n/a
B. Are security personnel included in the decisions of the BMP committee?	☐ yes ☐ no ☐ n/a
V. Is documentation of the security system filed separately from the BMP plan to prevent unauthorized individuals from obtaining confidential information?	☐ yes ☐ no ☐ n/a
EMPLOYEE TRAINING	
I. Is there an employee training program relative to the BMP program?	☐ yes ☐ no ☐ n/a

EMPLOYEE TRAINING (Continued)	
II. Can the employee training program, as part of the BMP plan, be incorporated into existing training programs?	☐ yes ☐ no ☐ n/a
III. Does the employee training program developed as part of the BMP plan focus on areas identified during the release identification and assessment stage as having the highest potential for harmful environmental releases?	☐ yes ☐ no ☐ n/a
IV. Will the employee training program adequately address changes resulting from the implementation of the BMP plan?	☐ yes ☐ no ☐ n/a
A. Have the audience and topics for the training been selected?	☐ yes ☐ no ☐ n/a
1. Have general sessions been planned to introduce the concept of pollution prevention, discuss the changes resulting from the BMP plan, and provide training in the new procedures?	☐ yes ☐ no ☐ n/a
2. Have separate training sessions been considered for specialized audiences (i.e., inspector, PM, and process specific training)?	☐ yes ☐ no ☐ n/a
3. Has appropriate speaker selection been taken into account to ensure effective training?	☐ yes ☐ no ☐ n/a
a. Has the expertise of persons outside the facility been utilized?	☐ yes ☐ no ☐ n/a
b. Are knowledgeable and enthusiastic speakers chosen?	☐ yes ☐ no ☐ n/a
B. Have materials been prepared to ensure the conduct of effective training?	☐ yes ☐ no ☐ n/a
1. Are technically accurate materials prepared for distribution at the training session?	☐ yes ☐ no ☐ n/a
a. Have references used in the development of this manual been consulted and utilized in developing training materials?	☐ yes ☐ no ☐ n/a
b. Are qualified personnel utilized in the development of materials?	☐ yes ☐ no ☐ n/a
2. Are eye-catching handouts and training tools (i.e., overheads, slides, videos) prepared?	☐ yes ☐ no ☐ n/a

EMPLOYEE TRAINING (Continued)	
3. Is ample time provided to develop training materials?	☐ yes ☐ no ☐ n/a
a. Are research time considerations taken into account?	☐ yes ☐ no ☐ n/a
b. Are speakers given ample time to prepare for the presentation, establish timing, and perform a practice presentation?	☐ yes ☐ no ☐ n/a
C. Are training events conducted to ensure that attendance is high and information is transferred in the most effective manner possible?	☐ yes ☐ no ☐ n/a
1. Are training events conducted in a positive manner with enthusiastic and knowledgeable presentations?	☐ yes ☐ no ☐ n/a
2. Was "hands-on" field training and employee participation incorporated where possible?	☐ yes ☐ no ☐ n/a
3. Are the schedules for the training sessions announced well in advance of the planned date?	☐ yes ☐ no ☐ n/a
4. Is training mandatory for all employees?	☐ yes ☐ no ☐ n/a
a. Does training include temporary or contractor personnel as well as permanent facility personnel?	☐ yes ☐ no ☐ n/a
b. Are new hires immediately instructed in BMPs?	☐ yes ☐ no ☐ n/a
5. Has outside assistance for the conduct of specialized presentations been considered?	☐ yes ☐ no ☐ n/a
D. Is training repeated when necessary?	☐ yes ☐ no ☐ n/a
1. Are meetings or training sessions held on a regular basis?	☐ yes ☐ no ☐ n/a
2. Is training repeated after facility changes are implemented which impact the BMP plan?	☐ yes ☐ no ☐ n/a
3. Are evaluation forms distributed at the end of each training session to determine both the effectiveness of sessions and the need for additional training?	☐ yes ☐ no ☐ n/a

RECORDKEEPING AND REPORTING	
I. Is there a recordkeeping and reporting program relative to the BMP program?	☐ yes ☐ no ☐ n/a
II. Can the recordkeeping and reporting program, as part of the BMP plan, be incorporated into established recordkeeping and reporting procedures?	☐ yes ☐ no ☐ n/a
III. Will the recordkeeping and reporting program ensure that records are appropriately kept and reporting is adequately conducted?	☐ yes ☐ no ☐ n/a
A. Are records developed in a standardized format?	☐ yes ☐ no ☐ n/a
1. Is there a standard format for submitting a report for internal review on accidental chemical releases or near-releases?	☐ yes ☐ no ☐ n/a
a. Does this report include adequate information such that educated decision can be mad (i.e., the area of release, volume of release, duration, and control measures and countermeasures used)?	☐ yes ☐ no ☐ n/a
2. Is there a standard format for reporting to the appropriate governmental regulating agency spills that reach the receiving water?	☐ yes ☐ no ☐ n/a
B. Does the BMP plan specify how information is to be transferred?	☐ yes ☐ no ☐ n/a
C. Is the identified method of communication effective?	☐ yes ☐ no ☐ n/a
1. Has verbal notification been considered to avoid a paperwork nightmare?	☐ yes ☐ no ☐ n/a
2. Are procedures adequate such that notification for accidental releases will be immediate?	☐ yes ☐ no ☐ n/a
a. Has a communication flow chart or other mechanism displaying the committee member's name, phone number, and responsibility been developed so that personnel will know precisely who needs to be notified in the event of a release?	☐ yes ☐ no ☐ n/a
b. Is a communication system (radio, telephone, public address system, or an alarm system) established for accidental chemical releases?	☐ yes ☐ no ☐ n/a

RECORDKEEPING AND REPORTING (Continued)	
i. Is the communication system affected by power outages?	☐ yes ☐ no ☐ n/a
ii. Is there a plan warning system that utilizes alarms to alert personnel of an unexpected release of material?	☐ yes ☐ no ☐ n/a
c. Do alarm systems, such as high-liquid-level alarms, for notification of impending spills adequately alert plant personnel?	☐ yes ☐ no ☐ n/a
d. Is the alarm system code posted and/or is it familiar to all plant personnel?	☐ yes ☐ no ☐ n/a
e. Are these alarms or signals displayed on a central control panel so that immediate communication to the supervisor or operator is achieved?	☐ yes ☐ no ☐ n/a
3. Have procedures been developed for notifying regulatory agencies of environmental releases?	☐ yes ☐ no ☐ n/a
a. Are personnel knowledgeable of responsibilities for reporting accidental chemical releases to regulatory agencies?	☐ yes ☐ no ☐ n/a
b. Are telephone numbers posted for the appropriate governmental regulating agencies (Federal, state, and local) which are to be notified in the event of a release of toxic and hazardous material to the receiving water?	☐ yes ☐ no ☐ n/a
c. Have designated personnel been identified to coordinate regulatory notification, thereby ensuring notification does occur?	☐ yes ☐ no ☐ n/a
4. Are personnel who will be receiving records and other notification aware of their responsibilities in reviewing and responding, if necessary, to the information?	☐ yes ☐ no ☐ n/a
a. Have review guidelines and response procedures been set forth?	☐ yes ☐ no ☐ n/a
b. Are the reviewing personnel in positions to require the prompt resolution of deficiencies found during the inspection?	☐ yes ☐ no ☐ n/a

RECORDKEEPING AND REPORTING (Continued)	
5. Are procedures adequate to ensure that the BMP committee is aware of the success of the BMP plan and any changes to or at the facility which warrant modification of the plan?	☐ yes ☐ no ☐ n/a
D. Has a filing system been developed to maintain the reports and records?	☐ yes ☐ no ☐ n/a
1. Have procedures been set forth to maintain records in an organized easily retrievable manner?	☐ yes ☐ no ☐ n/a
a. Are materials promptly filed?	
b. Are material filed in an organized manner?	☐ yes ☐ no ☐ n/a
2. Are records made, when appropriate, of verbal communication concerning the BMP plan?	☐ yes ☐ no ☐ n/a
3. Has one employee been chosen as the central recordkeeper to ensure that designated individuals review records where appropriate, and corrective action are identified and pursued?	☐ yes ☐ no ☐ n/a
BMP PLAN EVALUATION AND RE-EVALUATION	
I. Does the facility have a BMP plan in a narrative form?	☐ yes ☐ no ☐ n/a
II. Does the overall program appear to be comprehensive, understandable, and well organized?	☐ yes ☐ no ☐ n/a
A. Is the BMP plan readily available for review?	☐ yes ☐ no ☐ n/a
B: Does the BMP plan appear to address past, current, and potential environmental releases?	☐ yes ☐ no ☐ n/a
III. Is there a recognition of the need to periodically review and update the risk assessment and the facility's BMP plan as manufacturing conditions and applicable federal and state regulations change?	☐ yes ☐ no ☐ n/a
A. Is the BMP plan to be comprehensively reviewed at least once a year?	☐ yes ☐ no ☐ n/a
B. Is there a mechanism to keep the BMP committee apprised of changes at the facility?	☐ yes ☐ no ☐ n/a

BMP PLAN EVALUATION AND RE-EVALUATION (Continued)	
IV. Has a means of measuring the effectiveness of the BMP committee been considered (e.g., employee surveys, cost savings, analytical monitoring, reduced waste generation, etc.)?	☐ yes ☐ no ☐ n/a
V. Are facility-specific BMPs included in the plan?	☐ yes ☐ no ☐ n/a
A. Are the facility-specific BMPs in the program satisfactory?	☐ yes ☐ no ☐ n/a
VI. Has a system been devised to continue to evaluate plan components and incorporate modifications into the plan?	☐ yes ☐ no ☐ n/a

BMP COMMITTEE ROSTER		
Instructions: To help ensure that needed expertise is available and all responsibilities are considered, identify members of the BMP committee in the following table. List other needed specialists on the back.		
Leader:____ Title:____ Affiliation:____ Phone:____	Expertise:____	Duties:____
Member #1:____ Title:____ Affiliation:____ Phone:____	Expertise:____	Duties:____
Member #2:____ Title:____ Affiliation:____ Phone:____	Expertise:____	Duties:____
Member #3:____ Title:____ Affiliation:____ Phone:____	Expertise:____	Duties:____

BMP PLAN SUGGESTIONS

Instructions: Identify any problems/concerns and provide suggested improvements to plant operations which encompass facility-specific measures, and good housekeeping, preventive maintenance, inspections, security employee training, and recordkeeping and reporting programs. Be as specific as possible.

Suggestion #1:____________________

Suggestion #2:____________________

Suggestion #3:____________________

Suggestion #4:____________________

RELEASE POTENTIAL IDENTIFICATION AND ASSESSMENT WORKSHEET

		Identification and Assessment for Current Releases				Identification and Assessment for Potential Releases						
Steps 1 and 2		Steps 1 and 2	Step 3	Step 4	Step 5	Steps 1 and 2	Step 3				Step 4	Step 5
Process	Pollutants	Current Amount Released	Priority Based on Pollutant Toxicity	Priority Based on Pollutant Receptors	Priority Based on Current Amount Released	Potential Amount Released	Priority Based on Pollutant Toxicity	Current Controls	Confidence Rating of Controls	Priority Based on Potential Amount Released	Priority Based on Potential Pollutant Receptors	Priority Based on Potential Release

RECORDS TRACKING SHEET

Instructions: To help track records generated as part of the BMP plan, utilize the tracking sheet below.

Record Description	Date Received	Author	Date Routed	Recipient	Date Returned	Required Remedial Actions

APPENDIX C — THEORETICAL DECISION-MAKING PROCESS FOR BMP PLAN DEVELOPMENT

1.0 BACKGROUND

A small manufacturing company produces decorative hardware by forging, polishing, coating, and plating hardware sold for use in homes and businesses. This company operates two shifts per day, five days per week, and employs 65 persons. The company has been in operation since the early 1950s, and has updated the process equipment and the treatment systems twice, to increase cost effectiveness. This plant has experienced a history of National Pollutant discharge Elimination System (NPDES) permit limits exceedances of chromium and cyanide.

Through contacts within their trade association, other facilities, suppliers, and State and Environmental Protection Agency (EPA) inspectors, the owner and plant manager learned of other similar facilities that dramatically increased profitability through the implementation of best management practice (BMP) programs. The plant owner and manager believed that such increased profitability was beyond their capabilities since they believed it would encompass an expensive, overall plant modernization. Despite being incredulous, they investigated the possibilities of developing and implementing a similar plan.

The owner and plant manager worked together to gather information. They contacted State and EPA representatives to obtain pamphlets, case studies, and other documents which illustrated the benefits of pollution prevention. Additionally, they visited facilities similar to theirs which had implemented comprehensive environmental programs. Based on their review of the available information, they decided that the development of a BMP program would help to solve many of their environmental problems while proving to be profitable.

2.0 PLANNING PHASE

Much of the information reviewed by the owner and plant manager pointed to the need for a comprehensive environmental and management approach. A consistent theme was the importance of evaluating plant operations from the origination of pollution to the final disposition. Several manuals set forth steps to follow to reach this end. Generally, both agreed that the steps seemed reasonable and logical. As such, they formed a BMP committee, developed a policy statement, and performed a study to evaluate and prioritize current and potential discharges of pollutants.

2.1 FORMATION OF A BMP COMMITTEE

The owner and plant manager appointed themselves to be members of the committee because they were interested in making things happen; and since they were in charge, things would happen. They would certainly pay attention to the financial impacts of any potential changes. Since the plant processes were quite simple, they thought that at the most two more persons would be enough to form the committee. Several persons were approached about volunteering and few expressed interest including both shift foreman, and employee from the forging section who had voiced concerns about plant safety, a company receptionist, and a recently hired worker.

In evaluating these candidates the owner immediately accepted the employee who was concerned about safety issues. This decision was based on the need to both diversify input beyond management, and to gain a hands-on perspective for evaluating any changes they might consider. Past experience had shown that this employee was eager to express her opinions, and was very good at doing her job. These qualities seemed ideal.

The company receptionist and the new employee were considered but their lack of technical experience made them less likely candidates for committee members. However, the plant manager assured both persons that they would take and active role in the BMP plan implementation.

There was some hesitation about adding both shift foremen as this might be loading the committee with too many "chiefs"and not enough "indians". The owner and plant manager discussed the qualifications of both foremen and decided that the foreman with seniority was better qualified to act as the BMP committee member since he had been around when they tried to fine tune processes and waste handling in the past. The owner and plant manager felt that including this foreman, in lieu of the other, would make sure that past mistakes weren't repeated.

Having made this decision, the membership of the committee was set. The owner assigned the plant manager as the committee leader since he seemed the most plausible candidate. He then gave the plant manager the full reign is assigning activities and in moderating the committee.

The plant manager roughed out a schedule of activities and a list of responsibilities, and held a committee meeting to discuss BMP plan development. Generally, it was difficult to keep the committee focussed since all committee members seemed to have different agendas. Safety concerns, logistical problems, and financial considerations were all voiced. The plant manager assured everyone that evaluation of these problems would be addressed in the BMP plan and proceeded to introduce roles and responsibilities:

- The owner was assigned the task of developing the policy statement and informing employees of this initiative. The plant manager felt that the owner would be the most credible of the BMP committee members in formulating company policy.
- All four committee members were responsible with performing the release identification and assessment since this would involve a significant effort.
- The foreman was delegated primary responsibility of developing the good housekeeping program, with assistance from the plant manager. The plant manager believed that the foreman was in the primary position to assure that the good housekeeping program was implemented. Additionally, the plant manager felt that together, he and the foreman would be more inclined to make logistically sound decisions which would better assist them in meeting production schedules.
- The employee with safety concerns was assigned with developing the preventive maintenance and recordkeeping and reporting portions of the BMP plan. The plant manager felt that these programs involved and eye for detail and good organization skills, which had been demonstrated by this employee in the past.
- The owner felt obligated to address all security related issues since she had been responsible for addressing this area in the past and since security was somewhat outside the scope of any of the committee members' job descriptions.
- The plant manager assigned himself with the development of the inspection and employee training program. He felt that oversight of operations was primarily his responsibility and that he would best be able to evaluate their current inspection program. He also was familiar with all employees at the plant, and was currently in charge of the employee training program, thus making him the best candidate for integrating BMP plan training into the existing program.

The plant manager established a schedule for conducting weekly meetings to discuss plan development progress. He also provided the foreman and the employee with the information that had led to BMP plan development and development and suggested that they read the information prior to the next meeting. At the close of the meeting, the plant manager felt satisfied with the progress, but wished that he had provided the BMP plan-related documents to the committee members sooner, thus avoiding many of the questions that they had.

2.2 The BMP Policy Statement

The committee members decided to introduce the BMP program and the pollution prevention objectives by holding two meetings-one during each shift-which were open to all employees. The committee reasoned that this would help gain support from employees, and would make sure that employees were aware of the forthcoming changes and had a chance to take an active role in the company's success. They had developed a policy statement with some basic objectives, but felt that is was important to include employees in the development of the policy objectives. As this was a small facility, they decided to hold a ground meeting during each of the two shift's lunch hours. They offered free refreshments to increase voluntary attendance.

The owner created draft policy statement which was introduced in the group meetings. This statement read: "The objectives of the Best Management Practices Program at Sanchez Hardware are to reduce or eliminate pollutants in the wastewater discharges and increase profit." The meeting was facilitated by the owner. She spoke on what pollution prevention was all about and how pollution prevention activities could increase the profitability of the company such that everyone in the company would benefit with increased salaries, as well as company stability and growth. She described how some of the pollution prevention activities would change the way everyone ordinarily does their jobs. She then opened the meeting up for comments and questions.

The committee members were taken aback by the outpouring of concerns and ideas offered by the employees. The meetings had to be cut off due to tome limitations, but it was agreed that these meetings would be held again in the very near future. During these initial two meetings, employees expressed a number of concerns including:

- Two employees complained that working conditions were unsafe due to spills on the floors not being cleaned up immediately, as well as safety of some of the solvents and chemicals used in the plant. One person was concerned that the ozone layer may be depleted as a result of their solvent use.

- Some employees thought that time and money could be saved by relocating raw materials storage. It was considered to be too far away from the process area, and not set up in an organized fashion.

- Other employees were concerned that the so-called pollution prevention initiatives would make it hard for them to meet their quotas, and shipment dates, and that they weren't interested in changing operations.

The owner promised that the members of the BMP committee would address each of the issues raised at the open meetings. She requested that everyone feel free to provide information to the members of the committee regarding areas of the plant that should be closely evaluated for the need for pollution prevention activities. For this purpose, the owner discussed establishing a suggestion box in the employees' lounges.

After these meetings, the members of the BMP committee added to the policy statement to reflect the concerns of the employees. It became: "The objectives of the Pollution Prevention Program at Sanchez Hardware are to reduce or eliminate pollutants released to the environment, to increase profits, and to protect the worker's health and safety."

APPENDIX D — BIBLIOGRAPY

CHAPTER 2 REFERENCES

J. Cleary, J. Kehrberger, and C. Stuewe, "A Review of the Criteria for Evaluating a BMP Program," *Control of Hazardous Material Spills*, Proceedings of the 1980 National Conference on Control of Hazardous Material Spills, 1980.

NPDES Best Management Practices Guidance Document, EPA Industrial Environmental Research Laboratory, December 1979.

NPDES Best Management Practices Guidance Document, EPA Office of Water Enforcement and Permits, June 1981.

Waste Minimization Opportunity Assessment Manual (EPA 625 7-88 003), EPA Hazardous Waste Engineering Research Laboratory, July 1988.

C. Stuewe, J. Cleary, H. Thron, *Best Management Practices for Control of Toxic and Hazardous Materials*, Undated.

H. Thron, P. Rogoshewski, "Best Management Practices: Usefull Tools for Cleaning Up," 1982 Hazardous Materials Spills Conference, Undated.

Plant Maintenance Program Manual of Practice OM-3, Water Pollution Control Federation, Alexandria, VA, 1982.

Handbook for Using a Waste-Reduction Approach to Meet Aquatic Toxicity Limits, North Carolina Department of Environment, Health, and Natural Resources, Pollution Prevention Program, May 1991.

Storm Water Pollution Prevention for Industrial Activities, EPA Office of Wastewater Enforcement and Compliance, April 1992.

Facility Pollution Prevention Guide, EPA Office of Solid Waste and Risk Reduction Engineering Laboratory, Undated.

CHAPTER 3 REFERENCES

METAL FINISHING TEST REFERENCES

Industrial Pollution Prevention Opportunities for the 1990s (EPA 600 8-91 052), EPA Office of Research and Development, August 1991.

Development Document for Effluent Guidelines New Sources Performance Standards for the Metal Finishing Category, EPA Office of Water Regulation and Standards, June 1983.

METAL FINISHING TABLE REFERENCES

F1 *Development Document for Effluent Guidelines New Sources Performance Standards for the Metal Finighing Category*, EPA Office of Water Regulation and Standards, June 1983.

F2 G. Hunt, et al., *Accomplishments of North Carolina Industries-Case Summaries*, North Carolina Department of Resources and Community Development, January 1986.

F3 *Waste Reduction Assistance Program (WRAP) On-Site Consultation Audit Report: Electroplating Shop*, Alaska Health Project, April 1989.

F4 *The Robbins Company: Wastewater Treatment and Recovery System, A Case Study*, Office of Safe Waste Management, Massachusetts Department of Environmental Management, Undated.

F5 *A Study of Hazardous Waste Source Reduction and Recycling in Four Industry Groups in New Jersey, Case Study D6*, Jersey Hazardous Waste Facilities Siting Commission, April 1987.

F6 *Compendium on Low and Non-Waste Technology*, United Nations Economic and Social Counsel, 1991.

F7 *Guides to Pollution Prevention, The Fabricated Metal Industry (EPA 625/7-90/006)*, EPA Office of Research and Development, July 1990.

F8 *Metal Recovery: Dragout Reduction, Case History*, Minnesota Tehcnical Assistance Program, University of Minnesota, September 1988.

F9 D. Huisingh, L. Martin, H. Hilger, N. Seldman, *Proven Profits from Pollution Prevention: Case Studies in Resource Conservation and Waste Reduction*, Institute for Local Self-Reliance, Washington, D.C., 1985.

F10 R. Schecter, G. Hunt, *Case Summaries of Waste Reduction by Industries in the Southeast*, North Carolina Department of Natural Resources and Community Development, July 1989.

F11 [To be located]

F12 *Case Studies of Existing Treatment Applied to Hazardous Waste Banned from Landfill Phase II, Summary of Waste Minimization Case Study Results*, EPA Hazardous Waste Engineering Research Laboratory, October 1986.

F13 *Preliminary Report: Phase I Source Reduction Activities, Southeast Platers Project,* Massachusetts Department of Environmental Management Office of Safe Waste Management, July 1988.

F14 CALFRAN International, Inc., "Waste Reduction and Minimization by Cold Vaporization," *Process Technology '88, The Key to Waste Minimization, Volume 2,* August 15-18, 1988, held in Sacramento, California.

F15 *Wastestream Segregation, Recycling, and Treatment from an Electroplating Operation, Keysoton Plating, Hazardous Waste Minimization, A Resource Book for Industry,* San Diego County Department of Health Services, Undated.

F16 *Pollution Prevention Case Studies Compendium (EPA/600/R-92/046),* EPA Office of Research and Development, April 1992.

F17 *Waste Minimization Issues and Options Volume II (EPA/530-SW-86-04),* EPA Office of Solid Wasted and Emergency Response, October 1986.

F18 *Handbook for Using a Waste-Reduction Approach to Meet Aquatic Toxicity Limits,* North Carolina Department of Environment, Health, and Natural Resources Pollution Prevention Program, 1991.

F19 *Industrial Pollution Prevention Opportunities for the 1990s (EPA 600 8-91 052),* EPA Office of Research and Development, August 1991.

F20 *Addendum to Speakers' Notes,* The University of Tennessee Center for Industrial Services, March 13, 1991.

OCPSF MANUFACTURING TEXT REFERENCES

Industrial Pollution Prevention Opportunities for the 1990s (EPA 600 8-91 052), EPA Office of Research and Development, August 1991.

Development Document for Effluent Limitations Guidelines and Standards for the OCPSF Point Source Category (EPA 440/1-87/009), EPA Office of Water Regulations and Standards, October 1987.

Guides to Pollution Prevention, The Paint Manufacturing Industry (EPA 625/7-90/005), EPA Office of Research and Development, June 1990.

OCPSF MANUFACTURING TABLE REFERENCES

C1 G. Hunt, R. Schecter, *Accomplishments of North Carolina Industries-Case Studies,* North Carolina Department of Resources and Community Development, January 1986.

C2 *Process Technology and Flowsheets, Chemical Engineering*, McGraw Hill Publishing Company, New York, New York, 1979.

C3 *Pollution Prevention Guidance Manual for the Dye Manufacturing Industry*, EPA Office of Pollution Prevention, Undated.

C4 *Achievements in Source Reduction and Recycling for Ten Industries in the United States*, EPA Office of Research and Development, September 1991.

C5 D. Huisingh, L. Martin, H. Hilger, N. Seldman, *Proven Profits from Pollution Prevention: Case Studies in Resource Conservation and Waste Reduction*, Institute for Local Self-Reliance, Washington, D.C., 1985.

C6 D. Sarokin, W. Muir, C. Miller, S. Sperber, *Cutting Chemical Wastes: What 29 Organic Chemical Plants are Doing to Reduce Hazardous Wastes*, Inform, Inc., New York, New York, 1985.

C7 *Waste Minimization Issues and Options Volume II (EPA/530-SW-86-04)*, EPA Office of Solid Waste and Emergency Response, October 1986.

C8 *Handbook for Using a Waste-Reduction Approach to Meet Aquatic Toxicity Limits*, North Carolina Department of Environment, Health, and Natural Resources Pollution Prevention Program, 1991.

TEXTILE MANUFACTURING TEXT REFERENCES

G. Hunt, R. Schecter, *Accomplishments of North Carolina Industries-Case Studies*, North Carolina Department of Resources and Community Development, January 1986.

Recycling Zinc in Viscose Rayon Plants by Two-Stage Precipitation, EPA, Undated.

Industrial Pollution Prevention Opportunities for the 1990s (EPA 600 8-91 052), EPA Office of Research and Development, August 1991.

TEXTILE MANUFACTURING TABLE REFERENCES

T1 D. Huisingh, L. Martin, H. Hilger, N. Seldman, *Proven Profits from Pollution Prevention: Case Studies in Resource Conservation and Waste Reduction*, Institute for Local Self-Reliance, Washington, D.C., 1987.

T2 *Textile Oil Reclamation and Water Re-Use*, Osmonics, Inc., Undated.

T3 *Compendium on Low and Non-waste Technology*, United Nations Economic and Social Counsel, 1991.

T4 B. Smith, "Pollution Source Reduction (Part-II)" *American Dyestuff Reporter*, 1989.

T5 B. Handa, "Wastewater Management in a Synthetic Textile Industry", All India Workshop on Environmental Management of Small Scale Industries, July 22-23, 1989.

T6 H. Hiajue et al., "A Study of Reuse of Water in a Woolen Mill," Purdue University Conference on Industrial Waste Treatment, Undated.

T7 L. Paneerselvam, Director (PC), National Productivity Council, Lodhi Road, New Delhi 110 003.

T8 M. Sharma, Chief Chemist, Century Textiles and Industries Limited, Worli, Bombay 400 025, India.

T9 H. Asnes, "Reduction in Water Consumption in the Textile Industry," IFATCC Conference, London, 1978.

T10 S. Haribar, Senior Executive President, GRASIM, India, UNEP Workgroup, Paris.

T11 *Achievements in Source Reduction and Recycling for Ten Industries in the United States*, EPA Office of Research and Development, September 1991.

T12 *Waste Minimization Issues and Options Volume II (EPA/530-SW-86-04)*, EPA Office of Solid Waste and Emergency Response, October 1986.

T13 *Handbook for Using a Waste-Reduction Approach to Meet Aquatic Toxicity Limits*, Pollution Prevention Program of the North Carolina Department of Environment, Health, and Natural Resources, 1991.

T14 *Industrial Pollution Prevention Opportunities for the 1990s (EPA 600 8-91 052)*, EPA Office of Research and Development, August, 1991.

T15 *Waste Identification and Minimization: A Reference Guide*, 1987.

PULP AND PAPER MANUFACTURING TEXT REFERENCES

Estimates of Waste Generation by the Pulp and Paper Industry, Draft Report, EPA Office of Solid Waste, August 12, 1987.

The Product is the Poison: The Case for a Chlorine Phase-Out, Greenpeace, Washington, D.C., 1991.

USEPA/Paper Industry Cooperative Dioxin Study, The 104 Mill Study Summary Report, EPA Office of Water Regulations and Standards, July 1990.

Background Document to the Integrated Risk Assessment for Dioxins and Furans from Chlorine Bleaching in Pulp and Paper Mills (EPA 560/5-90-014), EPA Office of Toxic Substances, July 1990.

PULP AND PAPER MANUFACTURING TABLE REFERENCES

P1 *Compendium on Low- and Non-Waste Technology*, United Nations Economic and Social Counsel, 1991.

P2 *Pollution Prevention: Strategies for Paper Manufacturing*, University of Pittsburgh, Center for Hazardous Materials Research, Undated.

P3 "Memorandum: Strategy for the Regulation of Discharges of PHDDs and PHDFs from Pulp and Paper Mills to Waters of the United States," EPA Office of Water, May 21, 1990.

P4 *Summary of Technologies for the Control and Reduction of Chlorinated Organics from the Bleached Chemical Pulping Subcategories of the Pulp and Paper Industry*, EPA Office of Water Regulations and Standards and Office of Water Enforcement and Permits, April 1990.

P5 M. Sittig, *Pulp and Paper Manufacture, Energy Conservation and Pollution Prevention*, Noyes Data Corporation, Park Ridge, NJ, 1977.

P6 *Summary of Technologies for the Control and Reduction of Chlorinated Organics from the Bleached Chemical Pulping Subcategories of the Pulp and Paper Industry*, EPA Office of Water Regulations and Standards and Office of Water Enforcement and Permits, April 27, 1990.

PESTICIDES FORMULATION TEXT REFERENCES

Case Studies in Waste Minimization, Government Institutes, Inc., Rockville, MD, October 1991.

Industrial Pollution Prevention Opportunities for the 1990s (EPA 600 8-91 052), EPA Office of Research and Development, August 1991.

D. Sarokin, W. Muir, C. Miller, S. Sperber, *Cutting Chemical Wastes: What 29 Organic Chemical Plants are Doing to Reduce Hazardous Wastes*, Inform, Inc., New York, New York, 1985.

PESTICIDES FORMULATION TABLE REFERENCES

S1 *Case Studies in Waste Minimization*, Government Institutes, Inc., Rockville, MD, October 1991.

S2 "Fort Bliss' Hazardous Waste Minimization Plan," *Department of Defense Report on the Status of DOD Hazardous Waste Minimization*, March 31, 1988.

S3 *Guides to Pollution Prevention, The Pesticide Formulating Industry (EPA 626/7-90/004)*, EPA Risk Reduction Engineering Laboratory and Center for Environmental Research Information, February 1990.

S4 R. Schecter, G. Hunt, *Case Summaries of Waste Reduction by Industries in the Southeast*, North Carolina Department of Natural Resources and Community Development, July 1989.

S5 *Hazardous Waste Minimization Manual for Small Quantity Generators*, Center for Hazardous Materials Research, October 1989.

PHARMACEUTICAL MANUFACTURING TEXT REFERENCES

D. Sarokin, W. Muir, C. Miller, S. Sperber, *Cutting Chemical Wastes: What 29 Organic Chemical Plants are Doing to Reduce Hazardous Wastes*, Inform, Inc., New York, New York, 1985.

Industrial Pollution Prevention Opportunities for the 1990s (EPA 600 8-91 052), EPA Office of Research and Development, August 1991.

Guides to Pollution Prevention: the Pharmaceutical Industry (EPA 625/7-91/017), EPA Office of Research and Development, October 1992.

Preliminary Data Summary for the Pharmaceutical Manufacturing Point Source Category (440/1-89/084), EPA Waster Regulations and Standards, September 1989.

PHARMACEUTICAL MANUFACTURING TABLE REFERENCES

H1 *Guides to Pollution Prevention: the Pharmaceutical Industry (EPA 625/7-91/017)*, EPA Office of Research and Development, October 1992.

H2 D. Huisingh, L. Martin, H. Hilger, N. Seldman, *Proven Profits from Pollution Prevention: Case Studies in Resource Conservation and Waste Reduction*, Institute for Local Self-Reliance, Washington, D.C., 1985.

H3 *A Study of Hazardous Waste Reduction in Four Industrial Groups in New Jersey*, New Jersey Hazardous Waste Facilities Siting Commission, April 1987.

H4 [To be located]

H5 M. Melody, "Reducing the Waste in Wastewater", *Hazmat World*, August 1992.

PRIMARY METALS MANUFACTURING TEXT REFERENCES

Industrial Pollution Prevention Opportunities for the 1990s (EPA 600 8-91 052), EPA Office of Research and Development, August 1991.

Guidance Manual for Iron and Steel Manufacturing Pretreatment Standards, EPA Office of Water, September 1985.

PRIMARY METALS MANUFACTURING TABLE REFERENCES

M1 D. Huisingh, L. Martin, H. Hilger, N. Seldman, *Proven Profits from Pollution Prevention: Case Studies in Resource Conservation and Waste Reduction*, Institute for Local Self-Reliance, Washington, D.C., 1985.

M2 H. Nash, "Pretreatment and Recycle at Wire Rope Manufacture," Sixteenth Mid-Atlantic Industrial Waste Conference, Undated.

M3 *Catalogue of Successful Hazardous Waste Reduction/Recycling Projects*, Energy Pathways, Inc. and Pollution Probe Foundation, prepared for Industrial Programs Branch, Conservation and Protection Environment Canada, March 1987.

M4 "Recycling at California Steel Industries, Inc., Acid Wastes Become Profits." *Case Studies in Waste Minimization*, Government Institutes, Inc., Rockville, MD, October 1991.

M5 *Compendium on Low- and Non-Waste Technology*, United Nations Economic and Social Counsel, 1991.

M6 I. Rulkens, "Wastewater Problems in the Metal Industry: Results of Interviews in 48 Companies," TNO, Maatschappelijke Technologie, Postbus 342, 7300 AH, Apeldoorn, Netherlands.

M7 M. Stein, "Evaluation of the Chemelec Metal Recovery System as Applied to the Recovery of Zinc from Rinse Waters Following an Acid Pickle on a Barrel Zinc Plating Line," RIVM, Dept. LAE, Anthonie Van Leeuwenhoeklaan 1, Postbus 1, Bilthoven, Netherlands.

M8 M. Rachlitz, Vandig affedtning af stl inden malebehandling, Milig ~ styrelsen, 1990.

M9 [Source to be located]

M10 *Aluminum, Copper, and Nonferrous Metals Forming and Metal Powders Pretreatment Standards, A Guidance Manual*, EPA Office of Water, December 1989.

M11 *Development Document for Effluent Limitations Guidelines and Standards for the Aluminum Forming Point Source Category (EPA 440/1-82/073-b)*, EPA Office of Water and Waste Management, November 1982.

M12 *Guidance Manual for Iron and Steel Manufacturing Pretreatment Standards*, EPA Office of Water, September 1985.

M13 *Environmental Research Brief, Waste Minimization Assessment for an Aluminum Extrusions Manufacturer (EPA 600/S-92/018)*, EPA Risk Reduction Engineering Laboratory, April 1992.

M14 *Pollution Prevention Case Studies Compendium (EPA/600/R-92/046)*, EPA Office of Research and Development, April 1992.

M15 *New York State Waste Reduction Guidance Manual*, New York State Department of Environmental Conservation, March 1989.

PETROLEUM REFINING TEXT REFERENCES

The Generation and Management of Wastes and Secondary Materials: 1987-1988, American Petroleum Institute, Washington, D.C., June 1992.

Industrial Pollution Prevention Opportunities for the 1990s (EPA 600 8-91 052), EPA Office of Research and Development, August 1991.

Development Document for Effluent Limitations Guidelines and Standards for the Petroleum Refining Point Source Category (EPA 440/1-82/014), EPA Office of Water and Waste Management, October 1982.

Waste Minimization in the Petroleum Industry: A Compendium of Practices, American Petroleum Institute, Washington, D.C., November 1991.

PETROLEUM REFINING TABLE REFERENCES

R1 *Development Document for Effluent Limitations Guidelines and Standards for the Petroleum Refining Point Source Category (EPA 440/1-82/014)*, EPA Office of Water and Waste Management, October 1982.

R2 *Generation and Management of Wastes and Secondary Materials: Petroleum Refining Performance 1989 Survey*, American Petroleum Institute, Washington, D.C., June 1992.

R3 *Waste Minimization in the Petroleum Industry: A Compendium of Practices*, American Petroleum Institute, Washington, D.C., November 1991.

INORGANIC MANUFACTURING TEXT REFERENCES

Industrial Pollution Prevention Opportunities for the 1990s (EPA 600 8-91 052), EPA Office of Research and Development, August 1991.

Development Document for Effluent Limitations Guidelines and Standards for the Inorganics Chemical Manufacturing Point Source Category (EPA-440/1-80/007-b), EPA Office of Water and Waste Management, June 1980.

INORGANIC MANUFACTURING TABLE REFERENCES

N1 *Development Document for Effleunt Limitations Guidelines and Standards for the Inorganice Chemical Manufacturing Point Source Category (EPA-440/1-80/007-b)*, EPA Office of Water and Waste Management, June 1980.

N2 *Waste Minimization Issues and Options Volume II (EPA/530-SW-86-04)*, EPA Office of Solid Waste and Emergency Response, October 1986.

CHAPTER 4 REFERENCES

NTIS 1992 Catalog of Products and Services, U.S. Department of Commerce, Springfield, VA, October 1991.

Access EPA/IMSD-91-100, EPA Office of Administration and Resources Management, 1991.

M. Melody, R. McNulty, "Tap into Resources: Technical Assistance Programs Further Industry's Efforts", *Hazmat World, The Magazine for Environmental Management*, May 1992.

Pollution Prevention Resources and Training Opportunities in 1992 EPA/560/8-92-002, EPA Office of Pollution Prevention and Toxics and Office of Environmental Engineering and Technology Demonstration, January 1992.

ICPIC, International Cleaner Production Information Clearinghouse, United Nations Environment Programme in cooperation with EPA, Undated.

PPIC, Pollution Prevention Information Clearinghouse, EPA Office of Pollution Prevention and Office of Environmental Engineering and Technology Demonstration, April 1990.

CWRT WasteNotes, Center for Waste Reduction Technology, Summer 1992.

SWICH Memo, Solid Waste Information Clearinghouse, Undated.

Waste Reduction Institute for Training and Applications Research, WRITAR, Undated.

Environmental Law Handbook, 13th Edition — The recognized authority in the field, this invaluable text, written by 15 nationally-recognized legal experts, provides practical and current information on all major environmental areas. *Hardcover/550 pages/Apr '95/$79 ISBN: 0-86587-450-6*

Environmental Regulatory Glossary, 6th Edition — This glossary records and standardizes more than 4,000 terms, abbreviations and acronyms, all compiled directly from the environmental statutes or the U.S. Code of Federal Regulations. *Hardcover/544 pages/June '93/$68 ISBN: 0-86587-353-4*

Environmental Statutes, 1995 Edition — All the major environmental laws incorporated into one convenient source. *Hardcover/1,202 pages/Mar '95/$67 ISBN: 0-86587-451-4*
Softcover/1,202 pages/Mar '95/$57 ISBN: 0-86587-452-2
Also Available on Floppy Disk! *The following disks include* Folio© *Search and Retrieval Software.*
3.5" Floppy Disks for **DOS** *(#4054)* or **Windows** *(#4055)* ***$125***
Statutes Package, hardcover w/ ***DOS*** disk, *(#4056)* or w/**Windows** disk *(#4057)* ***$192***

Environmental Guide to the Internet — The new Environmental Guide to the Internet is your key to a wealth of electronic environmental information. From environmental engineering to hazardous waste compliance issues, you'll have no problem finding it with the easy-to-use **Environmental Guide to the Internet.**
Softcover/250 pages/Apr '95/$49 ISBN: 0-86587-449-2

Environmental Audits, 6th Edition — Details how to begin and manage a successful audit program for your facility. Use these checklists and sample procedures to identify your problem areas. *Softcover/592 pages/Nov '89/$79 ISBN: 0-86587-776-9*

Environmental Science and Technology Handbook — This book bridges the gap between the latest environmental science and technology available, and compliance with today's complex regulations. Contents include: Geology and Groundwater Hydrology; Air Quality; Environmental Processes; Human Health and Ecological Risk Assessment; Environmental Chemistry; Air Pollution Control Technologies; Solid and Hazardous Waste Treatment and Disposal; Underground and Aboveground Storage Tanks; Groundwater Pollution Control Technologies; and Pollution Prevention. *Softcover/389 pages/Dec '93/$79 ISBN: 0-86587-362-3*

Educational Programs

■ Our **COURSES** combine the legal, regulatory, technical, and management aspects of today's key environmental, safety and health issues — such as environmental laws and regulations, environmental management, pollution prevention, OSHA and many other topics. We bring together the leading authorities from industry, business and government to shed light on the problems and challenges you face each day. Please call our Education Department at (301) 921-2345 for more information!

■ Our **TRAINING CONSULTING GROUP** can help audit your ES&H training, develop an ES&H training plan, and customize on-site training courses. Our proven and successful ES&H training courses are customized to fit your organizational and industry needs. Your employees learn key environmental concepts and strategies at a convenient location for 30% of the cost to send them to non-customized, off-site courses. Please call our Training Consulting Group at (301) 921-2366 for more information!